CHEMICAL AND BIOCHEMICAL ENGINEERING

New Materials and Developed Components

AAP Research Notes on Chemical Engineering

CHEMICAL AND BIOCHEMICAL ENGINEERING

New Materials and Developed Components

Edited by
Ali Pourhashemi, PhD

Gennady E. Zaikov, DSc, and A. K. Haghi, PhD
Reviewers and Advisory Board Members

APPLE
ACADEMIC
PRESS

Apple Academic Press Inc. | Apple Academic Press Inc.
3333 Mistwell Crescent | 9 Spinnaker Way
Oakville, ON L6L 0A2 | Waretown, NJ 08758
Canada | USA

©2015 by Apple Academic Press, Inc.

First issued in paperback 2021

Exclusive worldwide distribution by CRC Press, a member of Taylor & Francis Group
No claim to original U.S. Government works

ISBN 13: 978-1-77463-346-5 (pbk)
ISBN 13: 978-1-77188-030-5 (hbk)

Library and Archives Canada Cataloguing in Publication

Chemical and biochemical engineering: new materials and developed components / edited by Ali Pourhashemi, PhD;
Gennady E. Zaikov, DSc, and A.K. Haghi, PhD Reviewers and Advisory Board Members.

(AAP research notes on chemical engineering)
Includes bibliographical references and index.
ISBN 978-1-77188-030-5 (bound)
1. Biochemical engineering. 2. Chemical engineering. I. Pourhashemi, Ali, editor II.
Series: AAP research notes on chemical engineering

TP248.3.C54 2015   660.6'3   C2014-908298-3

Library of Congress Cataloging-in-Publication Data

Chemical and biochemical engineering: new materials and developed components / Ali Pourhashemi, PhD, editor; Gennady E. Zaikov, DSc, and A.K. Haghi, PhD,

pages cm
Includes bibliographical references and index.
ISBN 978-1-77188-030-5 (alk. paper)
1. Biochemical engineering. 2. Chemical engineering. I. Pourhashemi, Ali, editor. II. Zaikov, G. E. (Gennadii Efremovich), 1935- III. Haghi, A. K.

TP248.3.C467 2015   660--dc23   2014047434

Apple Academic Press also publishes its books in a variety of electronic formats. Some content that appears in print may not be available in electronic format. For information about Apple Academic Press products, visit our website at **www.appleacademicpress.com** and the CRC Press website at **www.crcpress.com**

CHEMICAL AND BIOCHEMICAL ENGINEERING
New Materials and Developed Components

Edited by
Ali Pourhashemi, PhD

Gennady E. Zaikov, DSc, and A. K. Haghi, PhD
Reviewers and Advisory Board Members

Apple Academic Press Inc. | Apple Academic Press Inc.
3333 Mistwell Crescent | 9 Spinnaker Way
Oakville, ON L6L 0A2 | Waretown, NJ 08758
Canada | USA

©2015 by Apple Academic Press, Inc.

First issued in paperback 2021

Exclusive worldwide distribution by CRC Press, a member of Taylor & Francis Group
No claim to original U.S. Government works

ISBN 13: 978-1-77463-346-5 (pbk)
ISBN 13: 978-1-77188-030-5 (hbk)

Library and Archives Canada Cataloguing in Publication

Chemical and biochemical engineering: new materials and developed components / edited by Ali Pourhashemi, PhD;
Gennady E. Zaikov, DSc, and A.K. Haghi, PhD Reviewers and Advisory Board Members.

(AAP research notes on chemical engineering)
Includes bibliographical references and index.
ISBN 978-1-77188-030-5 (bound)
1. Biochemical engineering. 2. Chemical engineering. I. Pourhashemi, Ali, editor II.
Series: AAP research notes on chemical engineering

| TP248.3.C54 2015 | 660.6'3 | C2014-908298-3 |

Library of Congress Cataloging-in-Publication Data

Chemical and biochemical engineering: new materials and developed components / Ali Pourhashemi, PhD, editor; Gennady E. Zaikov, DSc, and A.K. Haghi, PhD,

pages cm
Includes bibliographical references and index.
ISBN 978-1-77188-030-5 (alk. paper)
1. Biochemical engineering. 2. Chemical engineering. I. Pourhashemi, Ali, editor. II.
Zaikov, G. E. (Gennadii Efremovich), 1935- III. Haghi, A. K.

| TP248.3.C467 2015 | 660--dc23 | 2014047434 |

Apple Academic Press also publishes its books in a variety of electronic formats. Some content that appears in print may not be available in electronic format. For information about Apple Academic Press products, visit our website at **www.appleacademicpress.com** and the CRC Press website at **www.crcpress.com**

ABOUT THE EDITOR

Ali Pourhashemi, PhD

Ali Pourhashemi, PhD, is currently a professor of chemical and biochemical engineering at Christian Brothers University (CBU) in Memphis, Tennessee, and has also taught at Howard University in Washington, DC. He taught various courses in chemical engineering, and his main area has been teaching the capstone process design as well as supervising industrial internship projects. He has published research articles in areas of heat transfer and packaging engineering and presented at many professional conferences. He is on the international editorial review board of the *International Journal of Chemoinformatics and Chemical Engineering* and is an editorial member of the *International Journal of Advanced Packaging Technology*. He is a member of the American Institute of Chemical Engineers.

REVIEWERS AND ADVISORY BOARD MEMBERS

Gennady E. Zaikov, DSc

Gennady E. Zaikov, DSc, is Head of the Polymer Division at the N. M. Emanuel Institute of Biochemical Physics, Russian Academy of Sciences, Moscow, Russia, and Professor at Moscow State Academy of Fine Chemical Technology, Russia, as well as Professor at Kazan National Research Technological University, Kazan, Russia. He is also a prolific author, researcher, and lecturer. He has received several awards for his work, including the the Russian Federation Scholarship for Outstanding Scientists. He has been a member of many professional organizations and on the editorial boards of many international science journals.

A. K. Haghi, PhD

A. K. Haghi, PhD, holds a BSc in urban and environmental engineering from University of North Carolina (USA); a MSc in mechanical engineering from North Carolina A&T State University (USA); a DEA in applied mechanics, acoustics and materials from Université de Technologie de Compiègne (France); and a PhD in engineering sciences from Université de Franche-Comté (France). He is the author and editor of 65 books as well as 1000 published papers in various journals and conference proceedings. Dr. Haghi has received several grants, consulted for a number of major corporations, and is a frequent speaker to national and international audiences. Since 1983, he served as a professor at several universities. He is currently Editor-in-Chief of the *International Journal of Chemoinformatics and Chemical Engineering* and *Polymers Research Journal* and on the editorial boards of many international journals. He is also a faculty member of University of Guilan (Iran) and a member of the Canadian Research and Development Center of Sciences and Cultures (CRDCSC), Montreal, Quebec, Canada.

AAP RESEARCH NOTES ON CHEMICAL ENGINEERING

The AAP Research Notes on Chemical Engineering series will report on research development in different fields for academic institutes and industrial sectors interested in advanced research books. The main objective of the AAP Research Notes series is to report research progress in the rapidly growing field of chemical engineering.

Ali Pourhashemi, PhD
Professor, Department of Chemical and Biochemical Engineering,
Christian Brothers University, Memphis, Tennessee, USA

Ing. Hans-Joachim Radusch, PhD
Polymer Engineering Center of Engineering Sciences,
Martin-Luther-Universität of Halle-Wittenberg, Germany

BOOKS IN THE AAP RESEARCH NOTES ON CHEMICAL ENGINEERING SERIES

Quantum-Chemical Calculations of Unique Molecular Systems
(2-volume set)
Editors: Vladimir A. Babkin, DSc, Gennady E. Zaikov, DSc, and
A. K. Haghi, PhD

Chemical and Biochemical Engineering
Editor: Ali Pourhashemi, PhD

Reviewers and editorial board members: Gennady E. Zaikov, DSc, and
A. K. Haghi, PhD

Clearing of Industrial Gas Emissions: Theory, Calculation, and Practice
Usmanova Regina Ravilevna, PhD, and Gennady E. Zaikov, DSc

CONTENTS

LIST OF CONTRIBUTORS

V. A. Babkin
Volgograd State Architect-build University, Sebrykov Department

I. I. Baholdin
Volgograd State Architect-build University, Sebrykov Department

Marina Bazunova
Bashkir State University 32 Zaki Validi Street, 450076 Ufa, Republic of Bashkortostan, Russia

Liliya Bazylyak
Physical Chemistry of Combustible Minerals Department, Institute of Physical-Organic Chemistry and Coal Chemistry named after L.M. Lytvynenko NAS of Ukraine 3aNaukova Str., Lviv, 79053, Ukraine

Sanjay Kumar Bharti
Institute of Pharmaceutical Sciences, Guru Ghasidas Vishwavidyalaya (GGV), Bilaspur, India; Tel: +91 7552-260027; Fax: +91 7752-260154; E-mail: sanjay.itbhu@gmail.com

A. A. Brilliant
GBUZSO Institute of Medical Cell Technologies, 620036 Yekaterinburg, Soboleva str., 25, Tel. +7(343)3769828

V. V. Chernova
Bashkir State University Russia, Republic of Bashkortostan, Ufa, 450074, ul. Zaki Validi, 32

A. A. Denisov
Volgograd State Architect-build University, Sebrykov Department

Sergey Gaydamaka
Moscow State University, Chemistry Faculty, Department of Chemical Enzymology. 119991, Moscow, Leninsky gory 1/11, Fax: +7-495-939-54-17

M. A. Kavrus
Grodno State Agrarian University, Grodno, Belarus

Azamat A. Khashirov
Kabardino-Balkarian State University a. Kh.M. Berbekov, 173 Chernyshevskogost., 360004, Nalchik, Russian Federation

Svetlana Yu. Khashirova
Kabardino-Balkarian State University a. Kh.M. Berbekov, 173 Chernyshevskogost., 360004, Nalchik, Russian Federation

Ol'ha Khlopyk
Physico-Mechanical Institute named after G.V. Karpenko NAS of Ukraine5 Naukova St., Lviv, 79060

O. L. Klyachenko
National University of Life and Environmental Sciences of Ukraine 03041, Kyiv, Heroiv Oboronu str., 15

E. I. Kolomiets
Institute of Microbiology, Belarus National Academy of Sciences, Minsk, Belarus

Ivan Krupenya
Bashkir State University 32 Zaki Validi Street, 450076 Ufa, Republic of Bashkortostan, Russia

S. A. Krylovska
National University of Life and Environmental Sciences of Ukraine 03041, Kyiv, Heroiv Oboronu str., 15

E. I. Kulish
Bashkir State University Russia, Republic of Bashkortostan, Ufa, 450074, ul. Zaki Validi, 32

Elena Kulish
Bashkir State University 32 Zaki Validi Street, 450076 Ufa, Republic of Bashkortostan, Russia

Andriy Kytsya
Physical Chemistry of Combustible Minerals Department, Institute of Physical-Organic Chemistry and Coal Chemistry named after L. M. Lytvynenko NAS of Ukraine 3aNaukova Str., Lviv, 79053, Ukraine

A. F. Likhanov
National University of Life and Environmental Sciences of Ukraine 03041, Kyiv, Heroiv Oboronu str., 15

Valentina Murygina Lomonosov
Moscow State University, Chemistry Faculty, Department of Chemical Enzymology. 119991, Moscow, Leninsky gory 1/11, Fax: +7-495-939-54-17

Debarshi Kar Mahapatra
School of Pharmaceutical Sciences, Guru Ghasidas Vishwavidyalaya (A Central University), Bilaspur, Chattisgarh, India

A. N. Michaluk
Grodno State Agrarian University, Grodno, Belarus

Vadim Z. Mingaleev
Institute of Organic Chemistry, Ufa Scientific Center of Russian Academy of Sciences, pr. Oktyabrya 71, Ufa, Bashkortostan, 450054, Russia

Vadim Z. Mingaleev
Institute of Organic Chemistry, Ufa Scientific Center of Russian Academy of Sciences, pr. Oktyabrya 71, Ufa, Bashkortostan, 450054, Russia

T. V. Romanovskaya
Institute of Microbiology, Belarus National Academy of Sciences, Minsk, Belarus

S. V. Sazonov
GBUZSO Institute of Medical Cell Technologies, 620036 Yekaterinburg, Soboleva str., 25, Tel. +7(343)3769828

A. S. Shurshina
Bashkir State University Russia, Republic of Bashkortostan, Ufa, 450074, ul. Zaki Validi, 32

Anamika Singh
Maitreyi Collage, University of Delhi

Rajeev Singh
Division of Reproductive and Child Health Indian Council of Medical Research, New Delhi; E-mail:
10rsingh@gmail.com

Sushil Kumar Singh
Department of Pharmaceutics, Indian Institute of Technology (BHU), Varanasi, India

N. V. Sverchkova
Institute of Microbiology, Belarus National Academy of Sciences, Minsk, Belarus

Bataeva Yulia
Federal State Budget Educational Institution of Higher Professional Education "Astrakhan State University." Home Address: 414 025, Astrakhan, st. Tatishchev, 16 E, fl. 307

V. P. Zaharov
Bashkir State University Russia, Republic of Bashkortostan, Ufa, 450074, ul. Zaki Validi, 32

Genadiy E. Zaikov
N.M. Emanuel Institute of Biochemical Physics of Russian Academy of Sciences, 4, Kosygin St.,
119991, Moscow, Russian Federation

Iriva D. Zakirova
Bashkir State University, Zaki Validi str. 32, Ufa, 450076 Bashkortostan, Russia

Gennadij Zaikov
Institute of Biochemical Physics named N.M. Emanuel of Russian Academy of Sciences 4 Kosygina
Street, 119334, Moscow, Russia

Gennadiy Zaikov
Institute of Biochemical Physics named after N. N. Emanuel, Russian Academy of Sciences 4 Kosygin
Str., 119991, Moscow, Russia

Vadim P. Zakharov
Bashkir State University, Zaki Validi str. 32, Ufa, 450076 Bashkortostan, Russia

Elena M. Zakharova
Institute of Organic Chemistry, Ufa Scientific Center of Russian Academy of Sciences, pr. Oktyabrya
71, Ufa, Bashkortostan, 450054, Russia

Y. M. Zasadkevich
GBUZSO Institute of Medical Cell Technologies, 620036 Yekaterinburg, Soboleva str., 25, Tel.
+7(343)3769828

N. S. Zaslavskaya
Institute of Microbiology, Belarus National Academy of Sciences, Minsk, Belarus

Azamat A. Zhansitov
Kabardino-Balkarian State University a. Kh.M. Berbekov, 173 Chernyshevskogost., 360004, Nalchik,
Russian Federation

LIST OF ABBREVIATIONS

AC	Active Sites
BD	Brownian Dynamics
BA	Butyl Acrylate
BMA	Butyl Methacrylate
CFT	Cefatoksim
CM	Continuum Modeling
CBC	Cyano-Bacterial Communities
DFT	Density Function Method
DAC	Dialdehyde Cellulose
DPD	Dissipative Particle Dynamics
DWNTS	Double Walled Carbon Nanotubes
DM	Dry Matter
ER	Estrogen
FEM	Finite Element Method
GC	Gas Chromatograph
GI	Gastrointestinal
GSDBT	Generalized Shear Deformation Beam Theory
GM	Gentamicin
HT	Heterotrophic Bacteria
HTS	High Temperature Shearing
HDB	Hybridoma Data Bank
HCO	Hydrocarbon Oxidizing Cells
HC	Hydrocarbons
IRIS	Immunogenetic Related Information Source
IFN	Interferons
LB	Lattice Boltzmann
LDPE	Low Density Polyethylene
MCC	Microcrystalline Cellulose
MWD	Molecular Weight Distribution
MC	Monte Carlo
MWNTS	Multi-Walled Nanotubes
MMA	Multifunctional Modifying Additive

NCM	Nano-Scale Continuum Modeling
NPS	Nanoparticles
CTZ	Polysaccharide Chitosan
PS	Polystyrene
PR	Progesterone
QM	Quantum Mechanics
QC	Quasi-Continuum
SWNTS	Single Walled Nanotubes
SEM	Spectroscopy
SAN	Styrene Copolymer with Acrylonitrile
TBMD	Tight Bonding Molecular Dynamics
TLR	Toll-Like Receptor
US	Ultrasound
UC	University of Cincinnati
VIM	Variational Iteration Method

LIST OF SYMBOLS

A_m^n and Y_m^n	the cross-sectional area and Young's modulus, respectively, of rod m of truss member type n
E_M	the modulus of the matrix
$E\rho$, E_θ, E_τ, E_ω	the energies associated with bond stretching, angle variation, torsion, and inversion, respectively
K_B	the Boltzmann constant.
$\{T\}$	the surface traction vector
v_f	the volume fraction of reinforcement
$[B]$	the matrix containing the derivation of the shape function
$\{d\}$	a vector containing the displacements
b_i	the effective rate constant of a fungus colony growth in the presence of the biocide
b_o	the effective rate constant of a fungus colony growth in the absence of the biocide
C	the biocide concentration
C	the stiffness tensor
D_{260}	absorbance of the solution at 260 nm
D_{280}	absorbance of the solution at 280 nm
E	the modulus of composite
$H(i)$ and $H(j)$	the *Hamiltonian* associated with the original and new configuration, respectively
k	a constant connected with parameters of interaction polymer-diffuse substance
K_c	a constant quantitatively equal to the biocide concentration
M	molecular weight
m_∞	relative amount of water in equilibrium swelling film sample
$m_{absorbed\ water}$	weight of the saturated condensed vapors of volatile liquid, g
m_{sample}	weight of dry sample, g
n	an indicator characterizing the mechanism of transfer of substance

pKa	universal index of acidity
q_{max}^{H+}	a maximum positive charge on atom of the hydrogen
ΔU	the change in the sum of the mixing energy and the chemical potential of the mixture
u, v and w	the displacement in x, y, z directions, respectively

GREEK SYMBOLS

$\bar{\sigma}_{fi}$	fibers average stress
$\bar{\sigma}_i$	composite average stress
$\bar{\sigma}_{mi}$	matrix average stress
β	the probability of chain termination
$\overline{F}_i(t)$	the force acting on the i-th atom or particle at time t
$\{\mathfrak{I}\}$	the force vector which contain both applied and body forces
$\psi(\beta)$	the distribution of active site over kinetic heterogeneity
Γ	the integration of the traction occurs only over the surface of the body
υ_0	the Poisson's ratio of the matrix

PREFACE

This book facilitates the study of problematic chemicals in such applications as chemical fate modeling, chemical process design, and experimental design. This volume provides comprehensive coverage of modern biochemical engineering, detailing the basic concepts underlying the behavior of bioprocesses as well as advances in bioprocess and biochemical engineering science. It combines contemporary engineering science with relevant biological concepts in a comprehensive introduction to biochemical engineering.

Cyanobacteria are known for their ability to make a significant contribution to soil fertility and enhance the growth processes of plants. In Chapter 1, laboratory experiments studied growth-stimulate activity of cyano-bacterial communities by using the test on the seeds of cress. In Chapter 2, a one-stage technique for the synthesis and modification of nanosized corrosion-inhibiting pigments based on zinc phosphate for the undercoatings with a good affinity to the organic phase has been developed. It was shown that the acrylic monomers can be used as the effective modifiers of the zinc phosphate nanoplates surface. The optimal concentration of modifying agent in the reactive medium is established. With the use of the electrochemical impedance spectroscopy, the anticorrosion activity of the obtained pigment in undercoating composition was investigated. It was determined that at the addition of 1% mass of the nanosized zinc phosphate, the efficiency of the undercoating is higher in comparison with the sample containing of 5% mass of the Novinox® PZ02 pigment.

Chapter 3 deals with development of a new aerobic-anaerobic bioremediation technology for impassable bogs polluted with oil in the north part of the Western Siberia; traditional remediation technologies are impossible technically and economically not favorable there. A research note on immunological databases and its role in immunological research is presented in Chapter 4. A simple scaffold with tremendous therapeutic potential is introduced in Chapter 5.

In Chapter 6 for the first time modification peculiarities of microcrystalline cellulose (MCC) and its oxidized form (dialdehyde cellulose DAC) of guanidine-containing monomers and polymers of vinyl and diallyl series have been studied, and the structure of the composites by IR spectroscopy and SEM has been researched. The biological activity of the synthesized composite materials was investigated and shown that the composite synthesized materials are quite active and have a biocidal effect against Gram-positive (St. Aureus) and Gram (E. coli) microorganisms.

A technical note on new nanocomposites based on layered aluminosilicate and guanidine containing polyelectrolytes is developed in Chapter 7.

Chapter 8 presents the results of experiments performed for sugar beet (*Beta vulgaris* L.) plant regenerants obtaining of one sort and five hybrids through the indirect morphogenesis. Stages and peculiarities of regenerants' indirect morphogenesis *in vitro* culture, as well as specific characters of morphogenic structures in sugar beet calluses, were studied. The effectiveness of triterpene saponosides usage as diagnostic markers, which determine potentially high productivity and adaptiveness of sugar beet (*Beta vulgaris* L.) plants, is proved.

Sorption and diffusive properties of films are studied in Chapter 9. Kinetic curves of release of medicinal substances having abnormal character are shown. The analysis of the obtained data showed that a reason for rejection of regularities of process of transport of medicinal substance from chitosan films from the classical fikovsky mechanism are structural changes in a polymer matrix, including owing to its chemical modification at interaction with medicinal substance.

A new way to increase activity of neodymium catalyst in the isoprene polymerization is shown in Chapter 10. There exists the possibility of increasing activity by hydrodynamic effect in turbulent flows at the stage of synthesis isopropanol complex with neodymium chloride, in this case the increase in the content of isopropanol in the complex and decrease the size of its particles. As a result the catalyst complex forms a stable in time and high activity when the polymerization of isoprene on the catalyst polymer has such a narrow MWD.

In Chapter 11, a research note on computational chemistry is presented.

The aim of Chapter 12 was elaboration of complex bacterial probiotic to facilitate biological accessibility of fodder, to promote immune correction and activation of metabolic processes in reared swine and poultry.

The aim of Chapter 13 is to investigate the polymerization of butadiene and isoprene in the presence of the catalytic system $TiCl_4$-$Al(iso$-$C_4H_9)_3$ under US irradiation of the reaction mixture during its formation.

The main objective of Chapter 14 is to investigate the interrelation between the particle size of a titanium catalyst and its kinetic heterogeneity in the polymerization of isoprene.

The aim of Chapter 15 is to evaluate modification of receptor profile of infiltrative breast carcinomas in accordance with dynamics of proliferative processes. The apportionment of the invasive carcinomas according to their proliferative activity was analyzed during the research. Dynamics of change of receptor status dependent on proliferative activity of the tumor was detected with the differentiated approach.

In Chapter 16, sorption properties of biodegradable polymer materials based on low-density polyethylene modified chitosans is studied.

In Chapter 17, it is shown that injection of silver nanoparticles (NPs) into large-tonnage polymers, such as polystyrene (PS) and styrene copolymer with acrylonitrile (SAN), imparts fungicide properties to them. For the purpose of further forecasting, the process of microscopic fungi and bacteria growth in the presence of different silver NP concentrations has been described quantitatively. It is shown that addition of silver NPs significantly suppresses growth of *Aspergillus niger* and *Penicillium chrysogenum* microscopic fungi both at the initial and stationary stages of their growth. Inhibition of bacterial growth manifests itself in increased induction period (the lag-phase).

New trends in carbon nanotube/polymer composites are presented in Chapter 18.

CHAPTER 1

INVESTIGATION OF PROPERTIES OF FITO-STIMULATE CYANO-BACTERIAL COMMUNITIES OBTAINED FROM LOWER VOLGA ECOSYSTEMS

BATAEVA YULIA

CONTENTS

ABSTRACT

Cyanobacteria are known for their ability to make a significant contribution to soil fertility and enhance the growth processes of plants. In laboratory experiments studied growth-stimulate activity of CBC by using the test on the seeds of cress.

1.1 INTRODUCTION

Here and now, agriculture is not the preferred chemical preparations, affecting negatively on soil fertility, environment, quality of products and biological agents that stimulate plant growth, yield environmentally friendly products and contribute significantly to soil fertility. These agents include the soil, rhizosphere, nitrogen-fixing bacteria that form numerous physiologically active substances that enter the roots of the plants and intensify the growth. They increase crop yields, reducing the ripening time, increase the nutritional value, improve resistance to disease, frost, drought and other adverse factors, speed up germination and rooting, reduce pre-harvest abscission of ovaries and abscission until the late frosts, are struggling with weeds and perform many other function.

A special place in the soil is occupied by cenoses algae and cyanobacteria. Cyanobacteria, unlike other soil algae fixed from the atmosphere, not only carbon but also atmospheric nitrogen, produce the biologically active substance and form of the primary production of organic matter [8–10]. Under natural conditions, cyanobacteria always develop in association with many other organisms, due to a mucous sheath, and, therefore, have excellent opportunities for adaptation and resistance to drastically changing physical and chemical environmental conditions. This creates the preconditions for more efficient appliances cyano-bacterial communities (CBC) at their introduction in the soil. In addition, cyanobacteria are economical to cultivation and have high growth rates, which is essential for the production of biologics. With rapid growth rates of cyanobacteria accumulate 20 days to 15 tons of biomass per 1 hectare.

In agricultural biotechnology cyanobacteria are poorly understood, not counting the rice fields [6]. The possibility of using cyanobacteria as fertilizers extensively studied in Asian countries, mainly in rice fields. Soil nitrogen-fixing cyanobacteria living cultures have a positive effect on growth and yield of rice [3, 4, 7].

The objective of our work was to study the properties of laboratory phyto-stimulate CBC from various aquatic and soil ecosystems of the Astrakhan region on the seeds of cress. To investigate the activity of phyto-stimulate used 25 laboratory collection of communities of cyanobacteria isolated from different aquatic and soil ecosystems of the Astrakhan region [1]. CLS collector supported by reseeding after 1–2 months on a liquid medium BG–11 in Erlenmeyer flasks 100–250 mL in volume and cultivation under natural light and a temperature of 22–25°C [8]. Cyanobacteria and algae are identified by morphological characters, using the determinant Gollerbah, etc. [2], textbook Zenov, Shtina [5].

Study of phytotoxicity and phyto-stimulate activity was performed using the test on the seeds of cress. For the experiment, the toxicity of the seeds of cress were placed in humid rooms sterile Petri dish with filter paper in triplicate. Pre-sterilized seeds treated with 70% ethanol for 3–5 min, then washed 3 to 5 times with sterile distilled water. Each camera was placed 50 seeds, which are wetted with slurry of 10 mL of sterile distilled water and 0.3 g biomass CBC pilot. The suspension was prepared by adding distilled water, chopped into small pieces of CBS strands (to reduce the gradient concentration of the suspension), then stirred for 3 min. Control seeds were soaked in sterile distilled water. Seeds treated with distilled water and suspensions were germinated for three days in daylight and 25°C.

Availability of growth-stimulate, inhibitory or neutral effect was determined by comparing the seed germination, root and stem length of plants in the control and experimental variants.

In studying the structure and composition of the investigated CBC found a large variety of habitats cyanobacteria, green and found diatoms. The main share of up representatives of cyanobacteria species of the genera: *Phormidium, Oscillatoria, Anabaena, Nostoc, Microcystis, Gloeocapsa* [1]. Fewer species are cyanobacteria genera *Chroococcus, Spirulina, Nostoc, Pleurocapsa, Synechococcus and Synechocystis*. Among the green algae is commonly encountered genus *Chlorella, Chlorococcum*.

1.2 RESULTS AND DISCUSSION

Results of the evaluation study of phyto-stimulate activity CBC presented in Table 1.1. As a result, data processing, were investigated CLS-toxic to the seeds of cress. The germination of seeds treated with CBC № 1, 2, 3, 6, 7, 8, 9, 10, 12, 13, 14, 20 was greater than in the control, equal to 85.7 ± 1.2%. Maximum germination of seeds 98.0 ± 1.5%, 95.0 ± 0.6%, 96.0 ± 1.4%, was observed in the processing of their communities of cyanobacteria № 7, 14, 20, respectively (Table 1.1).

TABLE 1.1 Effect of Bacterization of Cyano-Bacterial Communities in the Germination of Garden Cress

Variant (№ community of cyanobacteria)	Seed germination, %	The average size of the root length, mm	The average size of the length of the shoot, mm
Control	85.7±1.2	25.7±1.4	13.4±0.6
1	90.0±1.1	25.0±1.2	13.1±0.4
2	86.0±1.1	31.9±1.9	14.0±1.0
3	86.0±1.2	13.7±0.6	10.8±0.9
4	81.3±1.7	26.3±2.4	13.3±1.1
5	82.6±2.4	31.4±1.1	19.2±0.2
6	87.3±2.9	31.7±1.2	13.0±0.4
7	98.0±1.5	28.7±1.0	16.6±0.5
8	88.6±1.7	18.5±0.2	11.2±1.5
9	90.0±1.1	19.4±1.3	10.8±0.7
10	86.0±1.2	19.6±3.2	14.4±1.4
11	83.6±3.5	39.7±1.8	15.4±0.4
12	88.0±1.1	31.0±0.5	13.9±0.7
13	86.6±1.3	24.7±3.5	13.7±1.0
14	95.0±0.6	38.0±0.4	13.6±0.5
15	78.6±4.6	32.7±1.5	13.8±0.5
16	73.3±0.6	26.7±0.8	17.6±0.8
17	80.0±1.7	14.0±0.5	13.0±0.7
18	78.0±2.5	13.9±1.5	8.1±0.6

TABLE 1.1   *(Continued)*

Variant (№ community of cyanobacteria)	Seed germina- tion, %	The average size of the root length, mm	The average size of the length of the shoot, mm
19	82.0±2.1	16.9±0.5	11.4±1.4
20	96.0±1.4	23.6±0.5	12.0±1.5
21	84.0±1.1	43.7±0.2	17.5±0.3
22	82.0±1.2	15.9±1.5	14.4±1.6
23	82.0±1.2	18.0±1.2	14.4±1.5
24	64.8±3.1	4.1±1.2	8.0±0.6
25	84.0±1.8	15.8±1.8	12.1±0.7

Analysis of the data showed that the activity have growth-stimulate CBC № 2, 5, 6, 7, 11, 12, 14, 15, 16, 21. The inhibitory effect was observed in 12 CLS. The community number 24 there was a pronounced inhibitory effect; the average length of the root of 6 times and the average length of the stem is 1.6 times lower than in controls. Neutral effect was observed in seeds exposed bacterization CBC № 1, 4, 13. Some communities have shown stimulating activity relative to seedling root and stem growth suppressed, and vice versa (CBC № 6, 10).

The highest activity showed growth-stimulate community of cyanobacteria № 5, 11, 14 and 21. The average length of seedlings treated with CBC data, exceeded the control variant in the range of 5.8 to 18 mm. According to the literature it is known that growth-stimulate effect of cyanobacteria associated with the presence of auxin and gibberellin a like substances.

Thus, as a result of the experiment selected cyano-bacterial community № 2, 5, 6, 7, 11, 12, 14, 15, 20, 21, which can be used for further experiments, including field, with plants growing in the Astrakhan region, as well as for developments in agricultural biotechnology.

KEYWORDS

- Astrakhan region
- Cyanobacteria
- Cyano-bacterial community
- Phyto- and growth-stimulate activity
- Soil algae

REFERENCES

1. Bataeva, U. V., & Dzerzhinskay, I. S. (2010). Mwali Kamukvamba South of Russia: the environment development, 4, 76–78p.
2. Gollerbah, M. M., et al. (1953). Blue-green algae. Determinant of fresh water algae of the USSR, Moscow (in Russian).
3. Gollerbah, M. M., & Shtina, E. A. (1969). Soil Algae. Nauka, Leningrad, 228 p. (in Russian).
4. Goryunova, S. V., Rzhanova, G. N., & Orleanskiy, V. K. (1969). Blue-green algae (biochemistry, physiology, role in the practice). Nauka, Moscow (in Russian).
5. Zenova, G. M., & Shtina, E. A. (1990). Soil Algae: Tutorial. Moscow state university Moscow, 80 p (in Russian).
6. Pankratova, E. M., Trefilova, L. V., Zyablyh, R. Y., & Ustyuzhanin, I. A. (2008). Microbiology, 77(2), 266–272.
7. Pankratova, E. M. (1987). Success of Microbiology, 241–242.
8. Netrusov, A. I. (Ed). (2005). Workshop on Microbiology. Academia Moscow, 352 p. (in Russian).
9. Shtina, E. A., Zenova, G. M., & Manucharova, N. A. (1998). Soil Science. № 12.
10. Shtina, E. A., & Hollerbach, M. M. (1976). Nauka, Moscow 143 p.

CHAPTER 2

SYNTHESIS OF POLYMER-COATED ZINC PHOSPHATE NANOPARTICLES

LILIYA BAZYLYAK, ANDRIY KYTSYA, OL'HA KHLOPYK, and GENNADIY ZAIKOV

CONTENTS

ABSTRACT

A one-stage technique for the synthesis and modification of nanosized corrosion-inhibiting pigments based on zinc phosphate for the under coatings with a good affinity to the organic phase has been developed. It was shown, that the acrylic monomers can be used as the effective modifiers of the zinc phosphate nanoplates surface. The optimal concentration of modifying agent in the reactive medium is established. With the use of the electrochemical impedance spectroscopy the anticorrosion activity of the obtained pigment in undercoating composition was investigated. It was determined, that at the addition of 1% mass of the nanosized zinc phosphate the efficiency of the undercoating is higher in comparison with the sample containing of 5% mass of the Novinox ® PZ02 pigment.

PACS: 81.07.Bc; 82.35.Gh

2.1 INTRODUCTION

Modern demands to the ecological safety of production at the simultaneous decreasing of the unit value during the obtaining of the protective coating possessing by the anticorrosive properties cause the necessity to search the new ecologically pure and cheaper pigments. The phosphates (zinc phosphate and chromium phosphate) were the first compounds among those, which were used for the decreasing of the toxicity of the anticorrosive coatings instead of the widely applied chromium and stannum containing ones [1]. However, a general disadvantage of the phosphate pigments used in the anticorrosive coatings is a low efficiency of the under-film corrosion process evolution on the initial stages, which is connected with their low water solubility. Improving the properties of the anticorrosive coatings is possible through the use of nanosized pigments whose solubility is considerably higher than the common used. However, in references there is separate information about the methods of the zinc phosphates nanosized crystals synthesis, which in general, has the episodical character. Evidently, this is connected with the complication of the nanosized inorganic salts in aqueous solutions obtaining, which is explained by the fact that in spite of the very low solubility of zinc phosphate in water, in accordance with the Kelvin's equation the small crystals are dissolved and the big ones are aggregated. That is why the aim of the presented work was to find an effective method of mass transfer inhibiting during the synthesis of zinc phosphate nanoparticles.

2.2 EXPERIMENTAL PART

Butyl acrylate (BA) (Sigma-Aldrich, 99%), butyl methacrylate (BMA) (Sigma-Aldrich, 99%), Novinox ® PZ02 pigment (SNCZ, French Republic) and lacquer PF–170 (Yantar ®, Ukraine) were used as received.

Synthesis of zinc phosphate nanoparticles was carried out as described in Ref. [2, 3]. Form, size and the elementary composition of the synthesized product were estimated with the use of the scanning electron microscope EVO–40XVP (Carl Zeiss) with the system of the microanalysis INCA Energy 350 (Oxford Instruments). Anticorrosion effectiveness of the nanophosphate pigments in alkid coatings on D16T aluminum alloy were studied by electrochemical impedance spectroscopy using Gill AC potentiostat (ACM Instruments).

2.3 RESULTS AND DISCUSSION

Synthesis of zinc phosphate nanoparticles was carried in the presence of different concentration of acrylic monomers. BA and BMA were used as the modifying agents of the nanoparticles surface. The choice of such monomers was caused by fact that the BA and BMA is characterized by insignificant surface-active properties, and also it is practically insoluble in water; that is why at the new phase formation its molecules will be adsorbed on the surface of the zinc phosphate nanoparticles and will be prevented to their following aggregation. At the same time, under the final product drying-out at higher temperature the acrylates ion-coordinated polymerization process proceeding is possible and respectively, the formation of thin polymeric film on the surface of a pigment is possible; such film increases the affinity of the nanoparticles to the polymeric matrix of the paints and varnishes composition.

The influence of concentration of BMA in the reactive medium upon the size and composition of obtained pigment was studied (see Fig. 2.1).

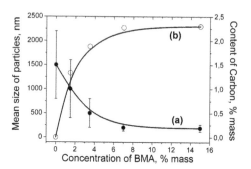

FIGURE 2.1   The dependence of the mean particles size (a) and the content of carbon in the pigments (b) on the concentration of bma in the reactive medium.

It was determined that the all products represent by themselves the thin plates with ratio between length and thickness equal to 10:1 (see Fig. 2.2a, b). The results of elementary analysis show that the ratio Zn:P in all samples is equal to 3 : 2.

FIGURE 2.2   SEM images of zinc phosphate nanoparticles obtained without modifying Agent (a), in the presence of BMA (b) and in the presence of BA (c).

As we can see from the Fig. 2.1, the optimal concentration of BMA for obtaining the nanoplates with the length 200 nm and thickness above 20 nm is 7% by mass. The presence of carbon in the samples confirms the assumption about the ion-coordinated polymerization of BMA on the nanoparticles surface.

In order to determine the influence of monomer's nature on the size and composition of phosphate pigments the synthesis of zinc phosphate in the presence of BA (7% by mass) has been done. It was determined (see Figs. 2.2c and 2.3c) that the pigment represent by themselves the thin plates with the mean length 400 nm and content 1.3% by mass of carbon.

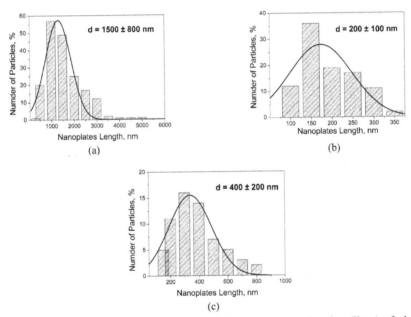

FIGURE 2.3   Size distribution (columns) and its gauss approximation (lines) of zinc phosphate nanoparticles obtained without modifying Agent (a), in the presence of 7% BMA (b) and in the presence of 7% BA (c).

The better results in a case of BMA, in comparison with BA, probably can be explained both considerably higher evaporation heat of BMA (near 75 cal/g for BMA and 46 cal/g for BA), and some better solubility of BA in water (near 0.08% for BMA and 0.2% for BA) [4]. The better solubility of BA in water may cause its worse adsorption on the nanoparticles surface than such of BMA. At the same time the high evaporation heat of BMA prevents the desorption of monomer during the drying and causes the formation of thickened polymer film on the nanoparticles surface than in the case of BA.

With the use of the electrochemical impedance spectroscopy the bilayer alkyd coatings based on the lacquer PF–170 superimposed on the duralumin alloy D16T was investigated. Such alloy has a much wide application in aircraft construction, transport, in building industry, etc. Typical protection scheme of the constructions in a case of such alloy application consists in the formation of the conversion layer on the surface of the metal, for example, at the expense of the oxidation, priming and upper varnish and paint layer.

In order to study the anticorrosive properties of the synthesized in the presence of BMA zinc phosphate nanoplates, the hinge of a pigment in quantity till 5% mass was added in priming layer of the alkyd coating PF–170 on the oxidized aluminum alloy. General thickness of the bilayer (inhibited coating + lacquer PF–170) coating was approximately 120 μm. In this same pentaphthalic coating the commercial zinc phosphate pigment Novinox ® PZ02 was used for comparison. In order to discover the protective effect from the inhibited components the defects by diameter 1 mm through the coatings have been done. It was determined that the both commercial and synthesized by us nanosized zinc phosphate essentially increase the resistance of the charge transfer of aluminum alloy with the defect alkyd coating in corrosive medium (Fig. 2.4). The interesting is fact, that the more anticorrosive effect is observed at the introduction into coating of the nanosized zinc phosphate in quantity from 1 till 3% mass, than 5% mass. Nanosized zinc phosphate is characterized by higher specific surface and that is why can possess by higher composition solubility and correspondingly, will be characterized by better-inhibited properties.

This indirectly means about the resistance decreasing for the corrosive solution contacting with the sample of coating having of zinc nanophosphate (Fig. 2.4). Alkyd coating with zinc nanophosphate at concentrations of 1 and 3% mass is characterized by the better anticorrosive protective properties than the coatings containing of 5% mass of the well–known zinc phosphate pigment Novinox ® PZ02. Such data point on the possibility to decrease the content of the inhibited pigment in alkyd coating in a case of the new synthesized by us nanosized pigment application on a basis of zinc phosphate.

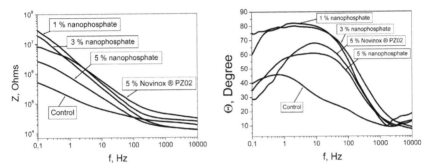

FIGURE 2.4 Frequency impedance dependencies of aluminum alloy D16T with inhibited alkyd coatings after 48 h in the medium of the acidic rain. Coatings contained the through defects by diameter 1 mm.

2.4 CONCLUSIONS

Nanosized inhibited pigments based on the zinc phosphate nanoparticles with the average linear sizes 200 nm and by the thickness 20 nm have been synthesized via ion change reaction in water medium. It was shown the possibility of acrylic monomers using for modification of the nanoparticles surface. It was determined, that the obtained nanoplates are characterized by better affinity to the organic solvent, in comparison with water. It was shown also, that the nanosized zinc phosphate is characterized by better inhibited properties in comparison with the well-known phosphate pigment Novinox ® PZ02 that permits to develop under industrial conditions the new competitive materials for their using in paint and varnish industry.

ACKNOWLEDGMENTS

This work has been performed under the partial financial support by grant № 4.11.1.42–14/K-2 via the framework "Nanotechnologies and Nanomaterials" from the National Academy of Sciences of Ukraine.

KEYWORDS

- Anticorrosive coating
- Butyl methacrylate
- Zinc phosphate nanoparticles

REFERENCES

1. Korsunsky, L. F., Kalinskaya, T. V., & Stiopin, S. N. (1992). *Inorganic Pigments*, Moscow: Khimiya, 336 p.
2. Pokhmursky, V., Kytsya, A., Zin,' I., Bazylyak, L., Korniy, S., & Hrynda, Yu. (25 March 2013). Patent of Ukraine №. 78529.
3. Pokhmursky, V., Zin,' I., Kytsya, A., Bily, L., Korniy, S., & Zin,' Ya. (25 March 2013). Patent of Ukraine № 78503.
4. Kargin, V. A. (1972). *Encyclopedia of Polymers (1),* Moscow: Sovietskaya Encyclopedia, 1224 p.

CHAPTER 3

SIMULATION IN THE LABORATORY CONDITIONS OF AEROBIC-ANAEROBIC BIOREMEDIATION OF OIL-POLLUTED PEAT FROM RAISED BOGS (RUSSIA)

SERGEY GAYDAMAKA and VALENTINA MURYGINA LOMONOSOV

CONTENTS

ABSTRACT

This chapter deals with development of a new aerobic-anaerobic bioremediation technology for impassable bogs polluted with oil in the North part of the Western Siberia, because traditional remediation technologies are impossible technically and economically not favorable there.

3.1   INTRODUCTION

The main oil production areas in Russia are situated in the Northern Siberia, and in the same places there are situated most extensive bogs polluted with oil. Application of remediation technologies, developed in Russia, on impassable bogs polluted with oil is almost impossible technically and economically unfavorable. Besides a severe climate with cold and long winters and short cool summers it is caused also by absence of any roads in tundra and forest-tundra and emergency oil spills on fenny bogs impassable for special machinery devices.

Therefore, an elimination of such spills and their consequences on the bogs is a very actual and difficult problem there. Depth of oil penetration on bogs doesn't exceed of 0.6–1.0 m and often is propped up with water or permafrost. Processes of self-restoration such bogs can prolong for several hundred years. The pollution can extend on width there and the irreparable damage will be caused to the Nature of the Polar Region.

In 2011, there was an attempt to clean from oil a strongly polluted bog with using (augmentation) of a bacterial oil-oxidizing preparation Rhoder. Oil spill was spring, and oil was partially collected with a pump for sludge. Three times the bog was watered with the Rhoder and one time with a fertilizer and lime. As a result level of oil pollution in the peat was decreased by 32%–98% depending on initial concentration of oil, which varied from 21–29 kg of crude oil on 1 kg of absolutely dry matter (DM) to 450–850 g/kg DM, and a depth of penetration of oil into the moss. The received results have induced a development of a new remediation technology in laboratory conditions with using of electron acceptors and the Rhoder to enhance of oil oxidation on the surface and in the depth of the peat. In this chapter there is presented an attempt to develop a new aerobic-anaerobic bioremediation technology for fenny bogs polluted with oil for using it in the Northern part of Russia.

3.2 MATERIALS AND METHODS

The microbial oil-oxidizing preparation Rhoder was used in laboratory experiment. The Rhoder consists of two bacterial strains Rhodococcus (R. ruber Ac–1513 D and R. erythropolis Ac–1514 D) picked out from soils, polluted with oil. Strains were not pathogenic for people, animals and plants and also don't cause mutations in bacteria. The Rhoder is allowed for broad use in the nature. It was successfully applied to bioremediation of oil sludge, soils, bogs and surfaces of water from oil pollution [1–8].

Installation, which was made from vertical plastic pipes (five models) with a length of 100 cm and diameter of 10 cm, was attached to a board (Fig. 3.1). In each model two openings with a diameter of 2 cm at distance of 40 cm and of 90 cm from the upper edge were made for sampling. Each model was filled with the natural peat polluted with oil with a high concentration of hydrocarbons (HC) from 370 g/kg to 550 g/kg of DM.

FIGURE 3.1   The laboratory installation, modeling a fenny bog, for development of a new technology of bioremediation of peat, polluted with oil.

3.2.1 SCHEME OF THE EXPERIMENT

- Model No. 1–negative control in which was added water for maintenance of high humidity of the peat, which was typical for bogs.
- Model No. 2–activation of indigenous microorganisms with mineral fertilizers that were added into the top layer of the peat into the depth of 10 cm, and introduction of a gaseous electron acceptor into the depth of 40 cm from the top layer of the peat in the model.
- Model No. 3–processing of the top layer of the peat into the depth of 10 cm with the water solution of the Rhoder and fertilizers and injection of the gaseous electron acceptor into the depth of 40 cm from the top layer of the peat in the model.
- Model No. 4–processing of the top layer of peat into the depth of 10 cm with the water solution of the Rhoder and fertilizers and liquid electron acceptor into the top layer of the peat in the model.
- Model No. 5–processing of the top layer of the peat into the depth of 10 cm with the water solution of the Rhoder and fertilizers and injection of the liquid electron acceptor into the depth of 40 cm from the top layer of the peat in the model.

3.2.1.1 CARRYING OUT BIOREMEDIATION

Soils in the models numbers 3–5 were processed three times with working solution of the Rhoder with a number of hydrocarbon oxidizing cells (HCO) of 1.0×10^8 cells/mL by watering with an interval in 3 weeks. The fertilizer ("Azofoska" C:N:P 16:16:16) was used, and 40 mL of the solution was added to the models three times. The gaseous and liquid electron acceptors were used and injected into models, according to the scheme of the experiment. The top layers of the peat in all models were maintained humidity not less than 60%, and the top layers of the peat were mixed two times a week and before each introduction of fertilizer and the Rhoder.

3.2.1.2 SAMPLING

Soils sampling from models were made before the experiment beginning and before every application of the fertilizer and the Rhoder and each injection of electron acceptors, which were entered into models. Samples from models were selected from the depth of 0–10 cm, 40 cm and 90 cm from the upper edge of each model for conducting of chemical, agrochemical and microbiological analyzes.

3.2.1.3   CHEMICAL AND AGROCHEMICAL ANALYZES

Oil in each sample of the dry peat was extracted on a Sockslet device with boiling $CHCl_3$, and gravimetrically determined. Then each dry material extracted by chloroform was fractioned on a mini-column with silica gel (Diapak-C). Oil products were analyzed by the gas chromatograph (GC). GC model is the KristalLuks 4000м (by company Metakhrom) with the NetChrom V2.1 program, the column OV-101 Length of 50 m, internal diameter of 0.22 mm, thickness of the phase of 0.50 microns, the FID detector, the temperature of the detector 300°C, the evaporator temperature of 280°C, the gradient from 80°C to 270°C, the velocity of raising temperature was 12°C per minute [9].

pH of each sample, humidity and the general maintenance of the available nitrogen and phosphorus were determined with colorimetric methods [10].

3.2.2   MICROBIOLOGICAL ANALYZES

MPN of microorganisms was determined by using tenfold dilutions and cultivations on meat-peptone agar in Petri dishes and using of selective agar nutrients for identification of ammonifying microorganisms, actinomicetes, pseudomonas, oligotrophic bacteria and micromycetes. MPN of anaerobic microorganisms (first of all SRB) in samples of the peat, which have been selected from the depth of 40 cm and 90 cm from models, were determined on the liquid Postgate's medium [11].

Determination of MPN of oil-oxidizing microorganisms in samples of peat were used the modified liquid Raymond's media with oil as a sole carbon source (g/L): Na_2CO_3-0.1; $CaCl_{2\times}6\ H_2O$-0.01; $MnSO_{4\times}7\ H_2O$-0.02; $FeSO_4$-0.01; $Na_2HPO_{4\times}12H_2O$-1.0; KH_2PO_4-1.0; $MgSO_{4\times}7\ H_2O$-0.2; NH_4Cl-2.0; $NaCl$-5.0; pH = 7.0 [12].

3.3   RESULTS AND DISCUSSION

Laboratory experiment was performed on the models which imitated of an over wetted bog polluted with oil. Preliminary microbiological analyzes of samples taken from the top, middle and bottom layers of the peat on the length of models showed that the peat in all models had different species of microorganisms: Bacillus, Pseudomonas, Rhodococcus, SRB and Penicillium. In the top layers of the peat (0–10 cm) in each model there were discovered the MPN of heterotrophic bacteria (HT) from 6.0×10^7 to 1.1×10^8 CFU/g of the peat, HCO bacteria from 9.1×10^5 to 9.4×10^6 cells/g of the peat. In the top layers of the peat in models anaerobic microorganisms didn't determine. In samples of the peat, which have been selected from the middle parts of models, the MPN of HT varied from 8.1×10^5 to 3.7×10^7 CFU /g of the peat, MPN of HCO bacteria varied from 8.2×10^2 to 9.8×10^4 cells/g of the peat. MPN of anaerobic microorganisms (SRB) 1.0×10^2 cells/g of peat were found in the samples from the middle of models Nos. 3 and 4. In the bottom samples of the peat in these models there were found anaerobic and microaerophilic bacteria with MPN from 2.1×10^6 to 4.9×10^7 CFU/g of the peat and HCO bacteria from 7.1×10^3 to 1.0×10^6 cells/g of the peat. SRB were found in the bottom samples from the models Nos. 2, 3 and 5 with the MPN from 1.0×10^2 to 1.0×10^5 cells/g of the peat (Table 3.1).

TABLE 3.1   Microbiological and Agrochemical Characteristics of Peat Samples from the Different Length of the Models before Bioremediation

№ model	Point of sampling	pH	HT CFU/g of peat	HCO, cells/g of peat	SRB, cells/mL	N-NH$_4^+$ mg/kg of peat	PO$_4^{3-}$ mg/kg of peat
1	Top	5.9	8.5×10^7	9.4×10^6	-	507.13	459.7
	Middle	6.2	8.1×10^5	8.1×10^2	0	516.2	393.6
	Bottom	6.7	2.9×10^6	9.7×10^4	0	544.1	418.3
2	Top	5.8	8.3×10^7	9.4×10^6	-	450.8	393.6
	Middle	6.4	3.6×10^7	9.8×10^4	0	453.7	471.1
	Bottom	6.1	1.4×10^7	7.1×10^3	$1,0 \times 10^5$	495.2	318.6

TABLE 3.1 *(Continued)*

№ model	Point of sampling	pH	HT CFU/g of peat	HCO, cells/g of peat	SRB, cells/mL	N-NH$_4^+$ mg/kg of peat	PO$_4^{3-}$ mg/kg of peat
3	Top	6.0	1.1×10^8	1.1×10^5	-	427.3	402.6
	Middle	6.2	3.7×10^7	8.7×10^4	1.0×10^2	440.7	401.1
	Bottom	6.3	4.9×10^7	1.1×10^4	$1,0 \times 10^2$	428.8	370.2
4	Top	6.3	1.1×10^8	9.1×10^4	-	402.5	391.2
	Middle	6.3	6.9×10^6	1.1×10^4	$1,0 \times 10^2$	448.6	411.9
	Bottom	6.1	2.8×10^6	9.4×10^4	**0**	463.9	506.7
5	Top	6.0	6.0×10^7	9.5×10^5	-	494.7	329.7
	Middle	6.2	3.4×10^7	1.1×10^4	0	454.1	278.1
	Bottom	5.9	2.1×10^6	1.0×10^6	$1,0 \times 10^5$	454.7	364.2

Agrochemical analyzes show that the content of nitrogen and phosphorus in the peat in models was rather high and the ratio of C:N:P was in average 100:0.1:0.01 (Table 3.1).

After completion of the experiment the MPN of microorganisms (HT and HCO bacteria) grew on 1 or 2–3 orders practically in all models in the top layers of the peat and decreased on 1–2 orders in the middle and bottom layers of the peat. At the same time the number of anaerobic microorganisms including SRB in all models significantly grew in the middle and the bottom layers of the peat that can be connected with formation of own biocenosis in each model (Table 3.2).

TABLE 3.2 Microbiological and Agrochemical Characteristics of Peat Samples from the Different Length of the Models after the End of the Bioremediation

№ model	Point of sampling	pH	HT CFU/g of peat	HCO, cells/g of peat	Other anaerobic bacteria/ SRB, cells/mL	N-NH$_4^+$ mg/kg of peat	PO$_4^{3-}$ mg/kg of peat
1	Top	5.8	5.7×10^7	1.1×10^7	-	316.9	168.9
	Middle	5.7	2.6×10^8	9.9×10^1	$0/5\times10^3$	425.8	276.7
	Bottom	5.2	3.4×10^7	8.2×10^3	$3.1\times10^8/$ 5×10^3	417.4	226.6
2	Top	6.3	2.1×10^9	8.6×10^7	-	493.7	429.5
	Middle	5.9	5.0×10^7	1.0×10^3	$2.0\times10^8/$ 5×10^3	297.9	273.5
	Bottom	6.5	8.0×10^7	6.1×10^6	$0/1.8\times10^4$	215.3	204.6
3	Top	6.4	1.0×10^9	8.8×10^7	-	589.9	173.5
	Middle	6.6	4.9×10^6	1.3×10^4	$1.2\times10^7/$ 5×10^3	589.4	369.9
	Bottom	6.3	4.7×10^6	1.1×10^6	$7.4\times10^7/$ 5×10^3	258.0	195.3
4	Top	6.8	2.2×10^8	1.0×10^8	-	229.1	251.3
	Middle	6.4	8.9×10^5	8.1×10^3	$6.6\times10^7/$ 5×10^3	214.0	203.2
	Bottom	6.5	1.2×10^6	8.5×10^3	$5.0\times10^7/$ 1.8×10^4	290.5	352.2
5	Top	6.5	4.3×10^8	1.0×10^8	-	122.5	349.8
	Middle	5.6	1.7×10^6	7.6×10^4	$1.5\times10^7/$ 1.8×10^4	313.3	142.9
	Bottom	6.2	1.2×10^6	1.2×10^3	$6.6\times10^7/5\times10^3$	294.7	199.2

Concentration of biogenic elements in all layers of models changed and even decreased that can be connected with activation of microorganisms in models. In the bottom and in the middle parts of the models concentration of biogenic elements decreased that can be connected with activation of anaerobic microorganisms in these models (Table 3.2).

Chemical analyzes have shown that the initial concentration of oil in the models have varied on height of the models from 370 g/kg DM to 550 g/kg DM. The concentration of oil in the models has decreased according the gravimetric analyzes on the average by 24–34%, including by 19% in the control model after finishing this experiment.

GC analyzes of oil products from models on their length are presented in Figs. 1–5.

Results of GC of the analyzes show that in control model there are processes degradation of oil in the top part. The quantity of peaks practically doesn't change, but their height and areas decreases (Fig. 3.2). The amount of oil products in the control model decreased in the top layer by 54%. In the middle of the model the number of peaks decreased by 1 peak, but their area increased. In the bottom of the model the quantity of peaks increased by 1 peak, but the area of peaks and their height (Fig. 3.2) significantly increased probably at the expense of an oil filtration down. GC analyzes of oil products on length of the model No. 2 are presented in Fig. 3.3.

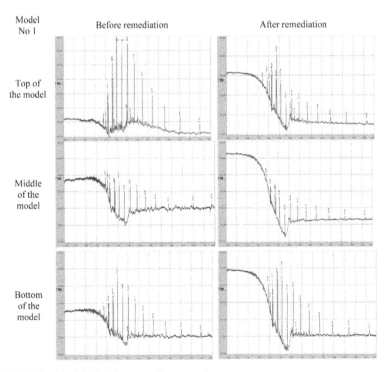

FIGURE 3.2   Model No 1 is a negative control.

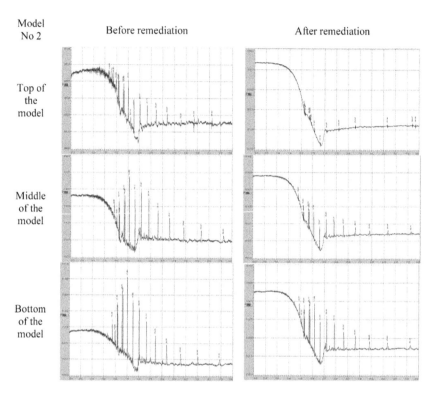

FIGURE 3.3 Model No 2 is the activation of indigenous microorganisms with fertilizers and injections of the gaseous electron acceptor into the middle part of the model.

Results of GC analyzes (Fig. 3.3) have shown that processes of oil degradation in the top, middle and bottom parts of this model took place. In the top and the middle parts of this model the quantity of peaks and their areas have decreased. In the bottom part of the model the quantity of peaks doesn't change, but their areas slightly have decreased. The concentration of oil products in the model decreased in the top part by 74%, in the middle part by 24% and in bottom by 5%. In this model the gaseous electron acceptor promotes degradation of oil products by anaerobic microorganisms in the middle and bottom parts. But in the top of the model aerobic indigenous microorganisms have worked. GC analysis of oil products in the model No.3 is presented in Fig. 3.4.

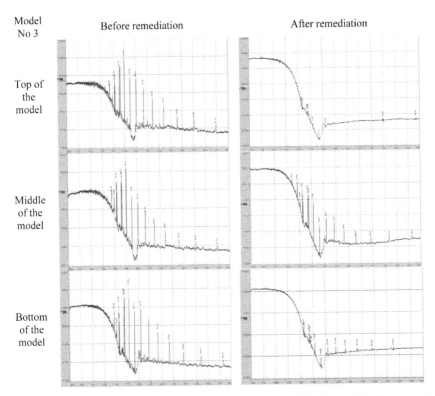

FIGURE 3.4 Model No. 3 is the augmentation with the Rhoder and fertilizers and injections of the gaseous acceptor of electrons into the middle part of the model.

Results of the GC analyzes (Fig. 3.4) have shown that in the model No 3, in which the Rhoder with the MPN of HCO bacteria 1.0×10^8 cells/mL and fertilizers were added three times and also the gaseous acceptor of electrons also were injected three times, have activated processes of oil degradation in the top and bottom parts of the model. In the middle part of the model the quantity of peaks and areas have increased. In the top part of the model the concentration of oil products have decreased by 88%, in bottom part by 68%. In the middle part of the model GC analyzes have not shown decrease in oil products. The acceptor of electrons probably has promoted degradation of oil products by anaerobic microorganisms in the bottom part of the model.

GC analyzes of oil products in the model No. 4 are presented in Fig. 3.5.

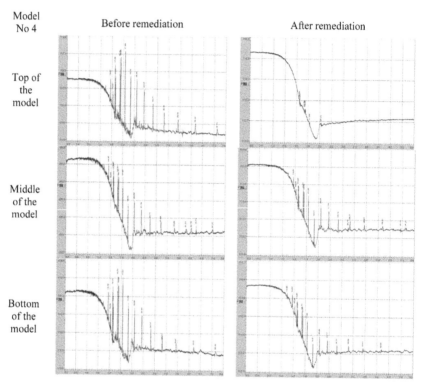

FIGURE 3.5   Model No. 4 is the addition of the Rhoder, fertilizers and the liquid acceptor of electrons into the top layer of the model three times.

Results of GC analyzes have shown (Fig. 3.5) that addition of the Rhoder, fertilizers and the liquid acceptor of electrons into the top layer of the model in three times has decreased the quantity of peaks from 15 to 6 and significantly has decreased its areas there. In the middle and the bottom parts of the model the quantity of peaks has not changed, but the areas of peaks in the bottom part of the model have decreased (Fig. 3.5). In this model the concentration of oil products have decreased in the top part almost by 96%, in the bottom part by 27%. In the middle part of the model the areas of peaks have increased by 24% by the end of the experiment. Probably the acceptor of electrons added into the top layer of the peat is not so good promoter for degradation of oil products by anaerobic microorganisms.

The GC analyzes of oil products in the model No. 5 are presented in Fig. 3.6.

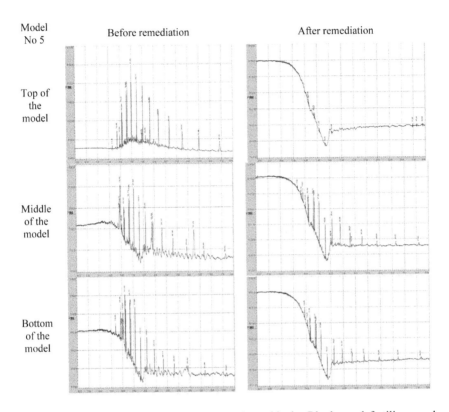

FIGURE 3.6 Model No. 5 is the augmentation with the Rhoder and fertilizers and injection of the liquid acceptor of electrons into the middle part of the model three times.

Results of GC analysis have shown that in the top part of the model (Fig. 3.6) the quantity of peaks has decreased from 18 to 10 and very significantly its total area has decreased. The concentration of oil products has decreased in the top layer of the peat by 99%. In the middle part of the model, the quantity of peaks doesn't change, but the total area of peaks has decreased. And concentration of oil products has decreased on the average by 64%. In the bottom part of the model the quantity of peaks has decreased from 19 to 15. The liquid acceptor of electrons, injected into the middle part of the model, significantly promotes degradation of oil products by anaerobic microorganisms in the middle and bottom parts in this model. In the top part of the model the process of oil degradation has provided with the Rhoder.

Thus, the received results show that the liquid electron acceptor in comparison with the gaseous electron acceptor has more good effect on the degradation of oil. And it is more expedient to inject the liquid electron acceptor into the middle part of the model (into the depth of 40 cm). These results are the first step to develop of a new aerobic and anaerobic bioremediation technology for strongly polluted fenny and almost impassable bogs in the North of Russia. Because there is impossible to collect completely spills of oil and perform a classic remediation technology on polluted bogs especially with using of specialized equipment and devices. It should be noted that the liquid acceptor of electrons has no relation to iron salts; this acceptor is ecofriendly and well makes activation of anaerobic indigenous microorganisms.

3.4 CONCLUSIONS

The obtained results showed that both studied acceptors of electrons well work into an anaerobic zone. Into the aerobic zone the Rhoder works more effectively in comparison with indigenous microorganisms. The oil oxidizing effect of the Rhoder in combination with the gaseous or the liquid acceptor of electrons showed good results. However augmentation with the Rhoder and fertilizer and the liquid electron acceptor, injected into the middle part of the model, had the best effect on oxidizing of oil there.

KEYWORDS

- Acceptor of electrons
- Augmentation
- Microorganisms
- Model
- Oil
- Peat

REFERENCES

1. Murygina, V. P., Arinbasarov, M. U., & Kalyuzhnyi, S. V. (1999). Ecology and Industry of Russia 8(16), (in Russian).
2. Murygina, V., Arinbasarov, M., & Kalyuzhnyi, S. (2000). Biodegradation 11(6), 385 (in Russian).
3. Valentina, P., Murygina, Maria, Y., Markarova, Sergey., & Kalyuzhnyi, V. (2005). Environmental International, 31(2), 163.
4. Ouyang, W., Yu, Y., Liu, H., Murygina, V., Kalyuzhnyi, S., & Xiu, Z. (2005). Process Biochemistry, 40(12), 3763.
5. Ouyang, Wei Liu., Hong, Yu., Yong-Yong., Murygina, V., Kalyuzhnyi, S., & Xiu, Zeng-De. (2006). Huanjing Kexue/ Environmental Science 27(1) 160.
6. De-Qing, S., Jian, Z., Zhao-Long, G., Jian, D., Tian-LI, W., Murygina, V., & Kalyuzhnyi, S. (2007). Water, Air, and Soil Pollution, 185(1–4), 177.
7. Murygina, Valentina, Markarova, Maria, & Kalyuzhnyi, Sergey. (2010). In Proc of IPY-OSC Symp Norway, Oslo: http://www.ipy-osc.no.
8. Murygina, V., Gaidamaka, S., Iankevich, M., & Tumasyanz, A. (2011). Progress in Environmental Science and Technology III 791.
9. Drugov, Yu. S., Zenkevich, I. G., & Rodin, A. A. (2005). Gas chromatography Identification of Air, water and Soil and Bio-nutrients Pollutants. Binom, Moscow, 752 p (in Russian).
10. Mineev, V. G., (ed.) (2001). Practical handbook on Agro chemistry, Moscow State University Moscow Russia, 688 p (in Russian).
11. Netrusov, A. I. (ed.) (2005). Practical handbook on microbiology Academia, Moscow Russia 608p (in Russian).
12. Nazina, T., Rozanova, Ye., Belyayev, S., & Ivanov, M. (1988). Chemical and microbiological research methods for reservoir liquids and cores of oil fields Preprint Biological Centre Press Pushchino, 35 p (in Russian).

CHAPTER 4

A RESEARCH NOTE ON IMMUNOLOGICAL DATABASES AND ITS ROLE IN IMMUNOLOGICAL RESEARCH

ANAMIKA SINGH and RAJEEV SINGH

CONTENTS

ABSTRACT

Immunological databases are the collection of information in a sequential and tabular manner, which help a user to access and retrieve the data related to immunology. In this chapter different aspects of immunological databases and its important role in immunological research is discussed.

4.1 INTRODUCTION

Immunological databases are growing day by day as the information related to disease is expanding tremendously. Nowadays Computational immunology expands itself and it is focused on analyzing large-scale experimental data and comparison [1, 2]. Immunology related databases cover all other aspects of immune system processes and diseases and the web address which are helpful for epitope designing and new drug designing [3].

Need of Databases development:
1. for the extraction of the existing information of diseases and immune related resources;
2. experiment designing on the basis of existing data;
3. analysis of experiments;
4. acceleration of knowledge based discovery;

4.2 DATABASE DEVELOPMENT

1. The most important aspect of immunology is immunological proteins which are large and they make difference by a single amino acid change and due to this single amino acid change there will be a significant change in the function of the proteins. Due to the databases it is easy to generate a comparative graph between two or more protein with similarity and differences.
2. To understand the origin, structure and function of antibodies MI–IC and other related immunological molecules.
3. To understand the mechanism behind immune disorders, infectious disease, autoimmunity, or tumor immunology.
4. Development and designing of new vaccines and antibodies.

At present large number of databases are available for applied and basic research in immunology. The databases are basically divided into two parts:

1. Sequence Databases
Collects the information of protein and DNA, RNA, etc. [4, 5].

2. Immunological Databases
Contains information of immune system related proteins and targets [6, 7].

4.3   DESCRIPTION OF DATABASES

A database is an organized collection of data. The data are typically organized to model relevant aspects of reality in a way that supports processes requiring this information. These general sequence databases are essential for molecular immunology projects because they provide interesting hits and useful insights about a particular sequence of immunological interest. For getting more information, there are many other specialized immunological databases. This lecture provides brief description about each immunological database.

4.3.1   ANTIGEN DB (HTTP://WWW.IMTECH.RES.IN/ RAGHAVA/ANTIGENDB/)

Sequence, structure, and other data on pathogen antigens [8].

4.3.2   IMGT (HTTP://WWW.EBI.AC.UK/IMGT)

It contains two databases, IMGT/LIGM–DB, a comprehensive database of Ig and TeR sequences from human and other vertebrates, and IMGT/HLA–DB, a database of human MHC. It enables users to extract data on nucleotide and protein sequences, sequence alignment, alleles, sequence tagged sites and polymorphisms, gene maps and genetic data, structural data, oligonueleotide primers, relationship with disease and cell lines [9].

4.3.3   FIMM (HTTP://SDMC.KRDL.ORG.SG.8080/GIMM)

FIMM focuses on cellular immunology, specifically on MHC, antigenic proteins, antigenic peptides, and relevant disease information. The tools include keyword search, pattern search, BLAST searches, multiple sequence alignment, and binding pocket/contact sites analysis [10].

4.3.4   VBASE (HTTP://WWW.MRC-CPE.CAM.AC.UK/IMT-DOC/ PUBLIC)

VBASE contains germ line variable region sequences of human antibodies. The search tool at this site helps to obtain amino acid and nucleotide sequences, scale maps of the human immunoglobulin loci, sequence alignments, numbers of functional segments, restriction enzyme cuts in V genes, and PCR primers for rearranged V genes [11].

4.3.5   HYBRIDOMA DATA BANK (HDB) (HTTP://WWW.ATCC. ORG/HDB/HDB.LITRNL)

This database contains information on hybridomas and other cloned cell lines and their immune reactive products (e.g., monoclonal antibodies). A HDB record contains comprehensive information on cloned cell lines including bibliography, biological origin, classification, methodological description, reactivity details, distributors, applications, availability and other relevant details [12].

4.3.6   MHCPEP (HTTP://BIO.DFCI.HARVARD.EDU/DFRMLI/)

This database contains list of MHC-binding peptides [13].

4.3.7   BCIPEP (HTTP://WWW.IMTECH.RES.IN/RAGHAVA/ BCIPEP)

The BCIPEP is a database of immunodominant peptides, which result in stimulation of B cell lineage. The database is consisting of nearly ~1100 B cell epitopes collected from literature. The data is kept in following fields: peptide sequence, antibody used for testing, reference, database reference of parent antigenic protein, Measure of antigenicity and immunogenicity [14].

4.3.8   SYFPEITHI (HTTP://WWW.UNI-TUEBINGEN.DE/UNI/KXI)

It is a database of MHC ligands and peptide motifs. Users can extract individual binding motifs and related peptides or search the database by peptide sequence. Additional options include search by anchor positions, peptide source, or peptide mass [15].

4.3.9   MHCDB (HTTP://WWW.LIMP.MRC.AC.UK/REGISTERED/ OPTION/MHEDB.HTML)

The database contain physical and genetic maps of human major histocom-patibility complex that include fully annotated genomic DNA sequences, cDNA sequences of class I and class II alleles [16].

4.3.10   HPTAA

(http://www.bioinfo.org.cn/hptaa/): HPTAA is a database of potential tu-mor-associated antigens that uses expression data from various expression platforms, including carefully chosen publicly available microarray ex-pression data, GEO SAGE data and Unigene expression data. [17].

4.3.11   HIV MOLECULAR IMMUNOLOGY DATABASE (HTTP:// HIV-WEB.LANL.GOV/IMMUNOLOV/INDEX.HTML)

It contains an annotated, searchable collection of HIV-1 cytotoxic and helper T-cell epitopes and antibody binding sites. The search tools include motif/pattern searches, sequence alignment to all sequences in the HIV-I genome and BLAST searches. The main aim of database is to provide comprehensive listing of HIV epitopes [18].

4.3.12   EPITOME (HTTPS://ROSTLAB.ORG/SERVICES/EPITOME/)

Epitome is a database of all known antigenic residues and the antibodies that interact with them, including a detailed description of the residues involved in the interaction and their sequence/structure environments. Each entry in the database describes one interaction between a residue on an antigenic protein and a residue on an antibody chain. Every interaction is described using the following parameters: PDB identifier, antigen chain ID PDB position of the antigenic residue, type of antigenic residue and its sequence environment, antigen residue secondary structure state, antigen residue solvent accessibility, antibody chain ID, type of antibody chain (heavy or light), CDR number, PDB position of the antibody residue, and type of antibody residue and its sequence environment. Additionally, interactions can be visualized using an interface to Jmol [19].

4.3.13   INTERFERON STIMULATED GENE DATABASE (HTTP:// WWW.LERNER.CCF.ORG/LABS/WILLIAMS/XCHIP-HTML.CGI)

Interferons (IFN) are a family of multifunctional cytokines that activate transcription of a subset of genes. The gene products induced by IFN are responsible for the antiviral, antiproliferative and immuno-modulatory properties of this cytokine. The database is fully searchable and contains links to sequence and Unigene information. The database and the array data are accessible via the World Wide Web.

4.3.14   MHCBN

The data of experimentally proven MHC binders, MHC nonbinders and T cell epitopes is a prime requirement for the development of prediction method for T cell epitope and/or MHC binders. The achievement of this goal is possible through the development of a comprehensive database in cellular immunology [20].

In past, number of databases has been created to provide information about MHC binding peptides and T cell epitopes. The databases like SY-FPEITHI, JenPep and HIV Database are modest in size and provide very focused information. Another database, FIMM contains information about MHC associated peptides, antigens, MHC molecules and associated disease. However, FIMM provides a rich set of internal/external data links and extraction of complex information, but it maintains only about 1500 naturally processed peptides or T cell epitopes. The MHCPEP is a widely used database that contains information about 13400 MHC binding peptides. It has greater proportional coverage than any of above-mentioned databases. The main limitations of MHCPEP are: i) it has not been updated since 1998; ii) database has no tools for data extraction/analysis; and iii) it is not linked with other database.

4.3.14.1   MHCBN TOOLS FOR EXTRACTION AND ANALYSIS OF DATA

The database has a set of web tools for extraction and analysis of data. The data extraction tools include **general query tool** and **peptide search tools** for making complex queries. The tools for the analysis of the data include:

i)   Tools for creation of datasets;
ii) MHC BLAST;
iii) Antigenic BLAST;
iv) Mapping of antigenic regions in query sequence; and
v)   Online submission of data.

The tool for mapping of antigenic regions is very useful tool for locating promiscuous antigenic regions in the query sequence.

4.4   PROJECTS BASED ON IMMUNOLOGICAL DATABASES

4.4.1   REFERENCE DATABASE OF IMMUNE CELLS (REFDIC)

RefDIC is an open resource of quantitative mRNA/Protein profile data specifically for immune cells [21, 22].

4.4.2   INNATE IMMUNE DATABASE (IIDB)

IIDB is a repository of computationally predicted transcription factor binding sites for over 2000 mouse genes associated with immune response behavior. A specific focus of IIDB is on Toll-like Receptor (TLR) genes, which are key components of innate immunity.

4.4.3   IMMUNOLOGY DATABASE AND ANALYSIS PORTAL (IMMPORT)

The ImmPort system provides information technology support in the production, analysis, archiving, and exchange of scientific data for researchers supported by NIAID/DAIT. It serves as a long-term, sustainable archive of data generated by investigators funded through the NIAID/DAIT. The Import system also provides data analysis tools and an immunology-focused ontology.

4.4.4   CASE STUDIES BASED ON IMMUNOLOGY DATABASE AND ANALYSIS PORTAL (IMPORT)

Study Title	PI	Type of Ex...	Public Rel...
⊟ Atopic Dermatitis & Vaccinia Network (ADVN) (14 Studies)			
SDY6: ADVN Biomarker Registry Study	Lisa Beck	ELISA	11/16/2012
SDY8: ADVN Biomarker Registry Study: CMI-HSV Substudy	Donald Leung	ELISPOT,EL...	11/16/2012
SDY7: ADVN Biomarker Registry Study: CMI/Ab-Vaccinia Substudy	Donald Leung	-	11/16/2012
SDY9: ADVN Biomarker Registry: Neutrophil Substudy	Lisa Beck	FCM,ELISA	11/16/2012
SDY5: Analysis and Correlation of Cathelicidin Expression in Skin and Saliva ...	Richard Gallo	-	11/16/2012
SDY13: Analysis of the Response of Subjects with Atopic Dermatitis to Oral ...	Donald Leu...	-	11/16/2012
SDY14: Antimicrobial Response to Oral Vitamin D3 in Patients with Psoriasis	Richard Gallo	-	11/16/2012
SDY4: Risk Factors in Atopic Dermatitis for the Development of Eczema Her...	Thomas Bie...	FCM	11/16/2012
SDY10: Role of Antimicrobial Peptides in Host Defense Against Vaccinia Virus	Donald Leung	-	11/16/2012
SDY11: Genetics of Atopic Dermatitis-Eczema Herpeticum	Lisa Beck, ...	-	
SDY12: Pilot Study to Determine the Underlying Mechanisms for Infection an...	Donald Leung	-	
SDY2: Immune Response to Varicella Vaccination in Subjects with Atopic D...	Lynda Sch...	FCM,ELISP...	11/16/2012
SDY3: Responses to Immunization with Keyhole Limpet Hemocyanin (KLH) ...	Henry Milgrom	-	11/16/2012

4.4.5   IMMUNOGENETIC RELATED INFORMATION SOURCE (IRIS)

IRIS is a database of all known human defense genes, produced in the laboratory of Professor John Trowsdale at the University of Cambridge. IRIS currently includes chromosomal locations, functional annotations, and sequence data for over 1,500 functional human immune genes. Please note that the IRIS database does not seem to be available anymore.

4.4.6 IMMUNOME DATABASE FOR GENES AND PROTEINS OF THE HUMAN IMMUNE SYSTEM

Immunome contains information about immune-related proteins, their domain structure and related ontology terms. Information can also be found for the localization of the coding genes and their comparison with the existing mouse orthologs [23, 24].

4.4.7 MACROPHAGES.COM

Macrophages.com is an online resource for those interested in macrophages and their role as major effector cells in innate and adaptive immunity. This website is designed to act as a centralized resource for the worldwide community of scientists interested in different aspects of macrophage biology [25].

4.5 PROJECTS BASED IMMUNOLOGICAL DATABASES

4.5.1 DC ATLAS PROJECT

DC-ATLAS is an immunological and bioinformatics integrated project, developed as a joint effort within the DC-THERA European Network of Excellence (www.dc-thera.org), a collaborating network established under the European Commissions Sixth Framework Programmed to translate discoveries from DC immunobiology into clinical therapies. The major scientific and technological goal of DC-ATLAS is to generate complete maps of the intracellular signaling pathways and regulatory networks that govern DC maturation/activation and function.

4.5.2 IMMUNOLOGICAL GENOME PROJECT (IMMGEN)

The Immunological Genome Project is a collaborative group of Immunologists and Computational Biologists who are generating, under carefully standardized conditions, a complete microarray dissection of gene expression and its regulation in the immune system of the mouse. The project encompasses the innate and adaptive immune systems, surveying all cell types of the myeloid and lymphoid lineages with a focus on primary cells directly ex vivo.

4.6   CONCLUSION

These databases provide information for understanding the specificity of immune system and immunological response at molecular level and designing a rapid in silico vaccine. Some of these databases provide special tools for in silico vaccine designing. The tools for detecting immunodominant region in an antigenic sequence are available at some databases, for example, antigenic mapping at MHCBN. The tools for identifying an immunological protein are available as BLAST against MHC database or antigenic protein database. Some database like MHC-PEP, MHCBN, FIMM, SYFPEITHI provide a collection of MHC binding peptides or T cell epitopes which form the basis of prediction methods for subunit vaccine design. Databases like SYFPEITHI, FIMM has tools for epitope prediction, which are used for novel epitope prediction. The databases which provide structural information about TCR, MHC and antigenic sequences are useful for making structure based prediction methods and rational drug designing. The blast tools available at various databases can be used to detect conserved epitopic region in different strains of pathogens.

KEYWORDS

- **FIMM**
- **Immunological databases**
- **MHCBN**
- **MHCPEP**
- **SYFPEITHI**

REFERENCES

1. Tong, J. C., & Ren, E. C. (July 2009). "Immuno informatics: current trends and future directions." Drug Discov, Today 14 (13–14), 684–689.

2. Korber, B., LaBute, M., & Yusim, K. (June 2006). "Immuno informatics comes of age." PLoS Comput. Biol 2(6): e71.
3. Ross, R. (1 February 1916). "An application of the theory of probabilities to the study of a priority pathometry Part I" (PDF). Proceedings of the Royal Society of London Series A 92 (638), 2042.
4. Sikic, K., & Carugo, O. (2010). "Protein sequence redundancy reduction: comparison of various methods." Bioinformation 5(6), 234–239. PMID 21364823.
5. Iliopoulos, I., Tsoka, S., Andrade, M. A., Enright, A. J., Carroll, M., Poullet, P., Prompo-nas, V., & Liakopoulos, T. et al., (Apr 2003). "Evaluation of annotation strategies us-ing an entire genome sequence" Bioinformatics 19 (6): 717–26.°PMID°12691983.
6. Tong, J. C., & Ren, E. C. (July 2009). "Immuno informatics: current trends and fu-ture directions." Drug Discov. Today 14 (13–14): 684–689. doi: 10.1016 /j.drudis. 2009.04.001 PMID 19379830.
7. Korber, B., LaBute, M., & Yusim, K. (June 2006). "Immuno informatics comes of age." PLoS Comput. Biol. 2(6), e71.
8. Ansari, H. R, Flower, D. R, & Raghava, G. P. (January 2010). "Antigen DB: an im-mune informatics database of pathogen antigens." Nucleic Acids Res. 38 (Database issue): D847–53. doi: 10.1093/nar/gkp830. PMC 2808902, PMID 19820110.
9. Lefranc, M. P. (January 2001). "IMGT, the international Immune Genetics data-base" Nucleic Acids Res. 29(1): 207–209. PMC 29797, PMID 11125093.
10. Schönbach, C., Koh, J. L., Flower, D. R, & Brusic, V. (2005). An update on the func-tional molecular immunology (FIMM) database Appl Bioinformatics, 4(1), 25–31.
11. Retter, I., Althaus, H. H., Münch, R., & Müller, W. (January 2005). "VBASE2, an integrative V gene database" Nucleic Acids Res 33 (Database issue) D671–674.
12. Rinsho, Nihon. (Nov 1992). 50(11), 2808–2815.
13. Zhang, G. L., Lin, H. H., Keskin, D. B., Reinherz, E. L., & Brusic, V. (2011). Dana-Farber repository for machine learning in immunology J Immunol Methods, 374(1–2): 18–25.
14. Saha, S., Bhasin, M., & Raghava, G. P. (2005). "Bcipep: a database of B-cell epit-opes." BMC Genomics 6, 79. doi: 10.1186/1471-2164-6-79.
15. Rammensee, H., Bachmann, J., Emmerich, N. P., Bachor, O. A., & Stevanović, S. (1999). SYFPEITHI: database for MHC ligands and peptide motifs. Immunogenetics, 50(3–4), 213–219.
16. Greenbaum, J. A., Andersen, P. H., Blythe, M., Bui, H. H., Cachau, R. E., Crowe, J., Davies, M., Kolaskar, A. S., Lund, O., Morrison, S., Mumey, B., Ofran, Y., Pellequer, J. L, Pinilla, C., Ponomarenko, J. V., Raghava, G. P., Regenmortel, M. H., Roggen, E. L., Sette, A., Schlessinger, A., Sollner, J. Z., & Peters, M. B. (2007). Towards a consensus on datasets and evaluation metrics for developing B-cell epitope prediction tools, J Mol Recognition, 20(2), 75–82.
17. Wang, X., Zhao, H., & Xu, Q. et al., (January 2006). "HP taa database-potential target genes for clinical diagnosis and immunotherapy of human carcinoma" Nucleic Acids Res. 34 (Database issue): D607–612.
18. Korber, B. T. M., Brander, C., Haynes, B. F., Koup, R., Moore, J. P., Walker, B. D., & Watkins, D. I. (2007). HIV molecular immunology, (2006)/(2007). Los Alamos, New Mexico: Los Alamos National Laboratory, Theoretical Biology and Biophysics.

19. Schlessinger, A., Ofran, Y., Yachdav, G., & Rost, B. (January 2006). "Epitome: database of structure-inferred antigenic epitopes." Nucleic Acids Res. 34 (Database issue): D777–380.

20. Bhasin, M., Singh, H., & Raghava, G. P. (March 2003). "MHCBN: a comprehensive database of MHC binding and nonbinding peptides." Bioinformatics 19(5): 665–666.

21. Hijikata, A., Kitamura, H., Kimura, Y., Yokoyama, R., Aiba, Y., Bao, Y., Fujita, S., Hase, K., Hori, S., Kanagawa, O., Kawamoto, H., Kawano, K., Koseki, H., Kubo, M., Kurita-Miki, A., Kurosaki, T., Masuda, K., Nakata, M., Oboki, K., Ohno, H., Okamoto, M., Okayama, Y., O-Wang, J., Saito, H., Saito, T., Sakuma, M., Sato, K., Seino, K., Setoguchi, R., Tamura, Y., Tanaka, M., Taniguchi, M., Taniuchi, I., Teng, A., Watanabe, T., Watarai, H., Yamasaki, S., & Ohara, O. (2007). Construction of an open-access database that integrates cross-reference information from the transcriptome and proteome of immune cells: *Bioinformatics*, 23, 2934–2941.

22. Kimura, Y., Yokoyama, R., Ishizu, Y., Nishigaki, T., Murahashi, Y., Hijikata, A., Kitamura, H., & Ohara, O. (2006). Construction of quantitative proteome reference maps of mouse spleen and lymph node based on two-dimensional gel electrophoresis, *Proteomics*, 6, 3833–3844.

23. Ortutay, C., Siermala, M., & Vihinen, M. Molecular characterization of the immune system: emergence of proteins, processes, and domains. Immunogenetics. doi: 10.1007/s00251-007-0191-0.

24. Ortutay, C., & Vihinen, M. (2007). Immunome: A reference set of genes and proteins for systems biology of the human immune system. Cell Immunol. doi: 10.1016/j.cellimm.01.01225.

25. Mossadegh-Keller, N., Sarrazin, S., Kandalla, P. K., Espinosa, L, Stanley, ER, Nutt, S. L., Moore, J., & Sieweke, M. H. (2013 May 9). M-CSF instructs myeloid lineage fate in single hematopoietic stem cells. Nature, 497(7448), 239–243.

CHAPTER 5

THIAZOLE: A SIMPLE SCAFFOLD WITH TREMENDOUS THERAPEUTIC POTENTIAL

SANJAY KUMAR BHARTI, DEBARSHI KAR MAHAPATRA, and SUSHIL KUMAR SINGH

CONTENTS

ABSTRACT

The treatment of infectious diseases still remains an important and challenging problem. Despite the availability of novel antimicrobial agents, still there is growing interest in this field. Many compounds have been synthesized with this aim but their clinical use has been limited by relatively high toxicity, bacterial resistance and/or lack of desired pharmacokinetic properties. A major research emphasis to counter this growing problem is the development of antimicrobials structurally unrelated to the existing molecules. One possibility to achieve this goal is the combination of two antimicrobial agents as a single molecule with appropriate biological activities.

5.1 INTRODUCTION

The heterocyclic compounds have received special attention as they belong to a class of compounds with proven physiological action. There are numerous biologically active molecules with five membered rings, containing two hetero atoms. Thiazole is an important scaffold known to be associated with several biological activities.

The occurrence of thiazole ring system in numerous biologically active molecules has been recognized which plays an important role in animal and plant kingdom. Different thiazole bearing compounds possess activities such as antibacterial [35], antifungal [39], anti-inflammatory [19], antihypertensive [32], anti-HIV [3], antitumor [26], anti filarial [27] anticonvulsant [29], herbicidal, insecticidal, schistosomicidal and anthelmintic [30]. The presence of thiazole ring in vitamin B_1 and its coenzyme play an important role as electron sink and for the decarboxylation of α-keto acids, respectively [8]. Many biologically active products, such as Bleomycin and Tiazofurin (antineoplastic agents) [31], Ritonavir (anti-HIV drug) [12], Fanetizole (N-phenethyl-4-phenylthiazol-2-amine), Fentiazac [(2-(4-(4-chlorophenyl)-2-phenylthiazol-5-yl)acetic acid)], Fenclozic acid ([2-(4-chlorophenyl)-1,3-thiazol-4-yl] acetic acid] and Meloxicam (anti-inflammatory agents) [28], Ravuconazole (4-(2-((2R,3R)-3-(2,4-difluorophenyl)-3-hydroxy-4-(1H-1,2,4-triazol-1-yl)butan-2-yl)thiazol-4-yl)benzonitrile) and Abafungin (antifungal agents), Nizatidine (antiulcer agent) [24], imidacloprid (insecticide) and penicillin (antibiotic) are some examples of thiazole bearing products. Thiazole derivatives are also widely used for the synthesis of antibiotic sulphathiazole [7], and with poly oxygenated phenyl component they showed promising antifungal activity. Thiazole nucleus as ligand of estrogen receptors [16] and also as novel class of antagonists for adenosine receptors [37] is known. Screening of 2, 4-disubstituted thiazoles as latent pharmacophores for diacylhydrazine of SC-51089, a potential PGE_2 antagonist have been reported [17]. The exciting results of 2,4-disubstituted thiazoles as a novel class of Src homology 2 (SH2) inhibitors for the treatment of osteoporosis and breast cancer have also been reported [9]. Synthesis of thiazole derivatives by various methods and their biological evaluation have been described by many researchers [18, 21, 14, 34, 22, 20, 11, 25, 15, 33]. Similarly, Schiff bases have gained importance because of physiological and pharmacological activities associated with them viz. antibacterial, antifungal, anticancer and antiviral agents, etc. [38]. Compounds containing azomethine group (–CH=N–) in the structure are known as Schiff bases, which are usually synthesized by the condensation of primary amines and active carbonyl groups. Since the thiazole moiety seems to be a possible pharmacophore in various pharmacologically active agents, we decided to synthesize compounds with this functionality coupled with Schiff base as possible antimicrobial and anti-inflammatory agents, which could furnish better therapeutic results (Fig. 5.1).

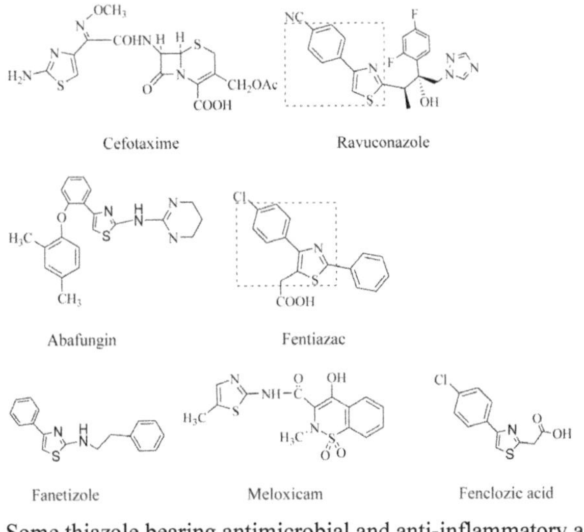

FIGURE 5.1   Some thiazole bearing antimicrobial and anti-inflammatory agents.

5.2 CHEMISTRY OF THIAZOLE

Thiazole is an important member of the azoles heterocycles containing nitrogen and a sulfur atom as part of the five-membered ring. Other members of azoles heterocycles include imidazoles and oxazoles. Thiazole ring is planar and aromatic in nature. It contains delocalized π electrons in the ring and the 6-π electrons satisfy Huckel's rule for its aromaticity. The π electron density makes C-5 as the primary site for electrophilic substitution, and C-2 as the site for nucleophilic substitution.

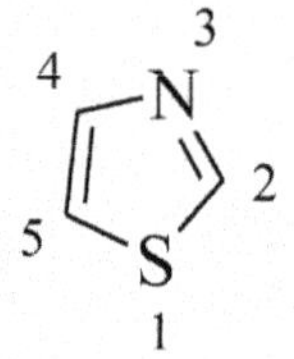

1,3-thiazole 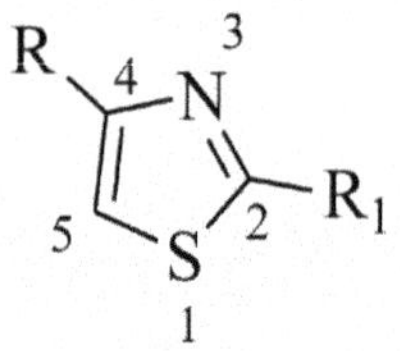   2,4-disubstituted thiazole

FIGURE 5.2   1,3-thiazole and 2,4-disubstituted thiazole

Thiazole can also be considered as functional group. Thiazole and related compounds are called 1,3–azoles. The resonance forms of thiazole are given in Fig. 5.3.

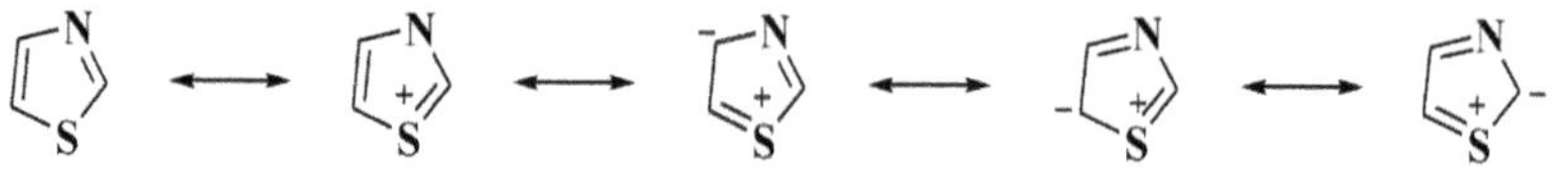

FIGURE 5.3   The resonance forms of thiazole.

Thiazole is a pale yellow liquid having pyridine like odor and molecular formula C_3H_3NS. It's boiling point is 116–118°C, specific gravity is 1.2, is sparingly miscible with water and miscible with alcohol and ether. It is used as an intermediate to manufacture synthetic drugs, fungicides and dyes.

Thiazole derivatives are used to control mold, blight, and other diseases caused by fungus in fruits and vegetables. It is also used as a prophylactic treatment for Dutch elm disease. Its use in treatment of diseases caused by Aspergillus has been reported. As an antiparasitic, it is able to control roundworms, hookworms, and other helminth species which attack wild animals, livestock and human. Its mode of action is thought to be due to inhibition of the specific helminth's mitochondrial enzyme, fumarate reductase, with possible interaction with endogenous quinone etc. Related compound thiabendazole is used in the treatment of metal poisoning, such as lead poisoning, mercury poisoning or antimony poisoning. It is also used as a food additive, as preservative and to treat ear infection of animals.

The thiazolothiazoles are in greatest demand because they are used in optical recording materials, as lightness accelerators, in silver-halide photographic materials and also as semiconductors.

Pyrazolopyrazoles are used as antiviral agents and in the treatment of mucolipidosis type I (ML I) or sialidosis an inherited lysosomal storage disease that results from a deficiency of the enzyme sialidase. The lack of this enzyme results in an abnormal accumulation of complex carbohydrates, mucopolysaccharides and fatty substance mucolipids. Both of these substances accumulate in body tissues.

Thienothiazoles are used as spectral sensitizers for photographic materials and silver-halide emulsions. Their derivatives are used as sensitizers in photo films for medical radiograms. Pyrrolothiazoles have shown anti-inflammatory and immune modulatory effects. They are also used as anticoagulants, inhibitors of factor Xa, and for the prevention and treatment of thrombosis and embolism. Utility component of photographic films and papers has also been reported.

The thiazole systems condensed with six-membered hetero aromatic rings have also been reported. Derivatives of thiazolopyridine exhibit various biological activities including analgesic, antipyretic, anti-inflammatory, and antimicrobial activities. They are also used as therapeutic agents for the treatment of diabetes, sexual dysfunction and hyper proliferative disorders (Fig. 5.4).

Among the different aromatic heterocycles, thiazoles occupy a prominent position in the drug discovery process and this ring structure is found in several natural and marketed products. Numerous natural products containing thiazole nucleus have been isolated and exhibited significant biological activities such as cytotoxic, immunosuppressive, antifungal, and enzyme inhibitory activities. Recently, great attention has been paid to design, synthesize and evaluate biological activity of compounds with thiazole ring system as pharmacophore. For example, the thiazole bearing epothilones (A–F) have been reported as a new class of anticancer drugs (Fig. 5.5). Early studies in cancer cell lines and in human cancer patients indicated better efficacy and milder adverse effects than taxanes. They prevent cancer cells from dividing by interfering with tubulin protein similarly as taxane. Their mechanism of action is similar, but their chemical structure is different containing thiazole ring. Due to their better water solubility, cremophors (solubilizing agents used for paclitaxel, which can affect cardiac function and cause severe hypersensitivity) are not needed. Endotoxin like properties known from paclitaxel, activation of macrophages, inflammatory cytokines and nitric oxide, are not observed for Epothilone B. Epothilones were originally identified as metabolites produced by the mycobacterium *Sporangium cellulosum*.

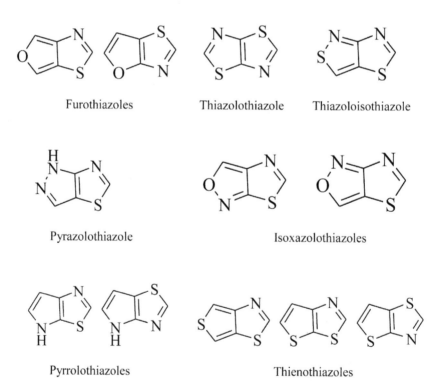

FIGURE 5.4 Thiazole systems condensed with five-membered heteroaromatic rings.

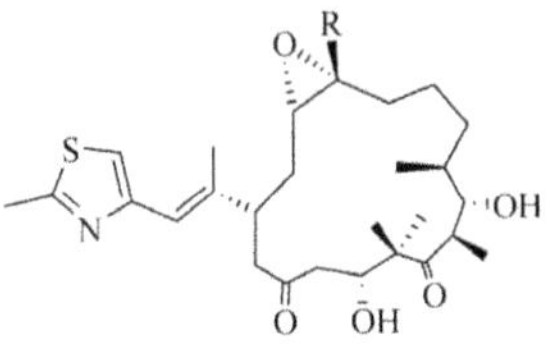

Epothilones A (R = H) and B (R = CH₃)

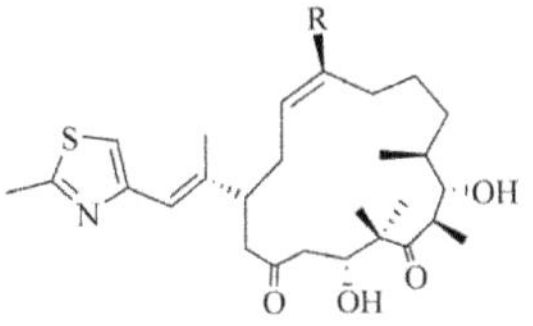

Epothilones C (R = H) and D (R = CH₃)

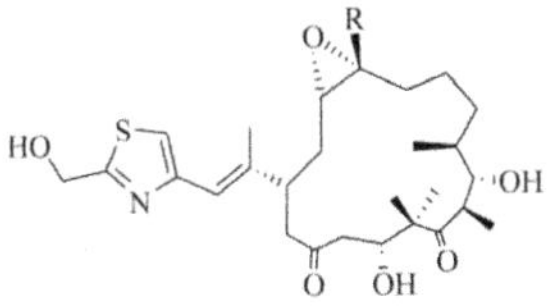

Epothilones E (R = H) and F (R = CH₃)

FIGURE 5.5   Epothilones A–E.

The algaecidal activity of Bacillamide A and two new analogs Bacillamides B and C (Fig. 5.6) containing thiazole nucleus in their structures have been isolated from *Bacillus endophyticus*.

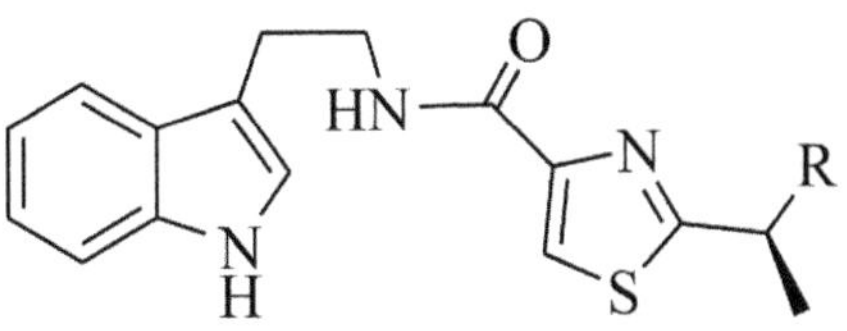

Bacillamide A: R= O

Bacillamide B: R= OH

Bacillamide C: R= NHAc

FIGURE 5.6   Bacillamides A, B and C.

A related compound, Neobacillamide A (Fig. 5.7) together with Bacillamide C, has been isolated from the bacterium *Bacillus vallismortis* and evaluated for their cytotoxic activity (Yu et al., 2009). Both Neobacilla-mide A and Bacillamide C were found inactive as cytotoxic against HL60 human leukemia cells and A549 human lung cancer cells.

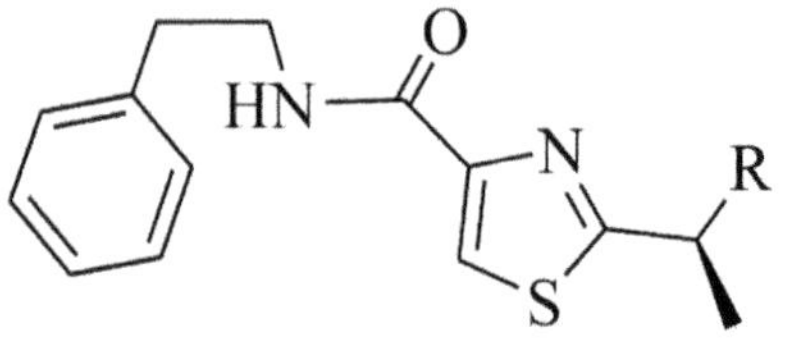

FIGURE 5.7   Neobacillamide A.

A thiopeptide antibiotic, Urukthapelstatin A (Fig. 5.8) was isolated from a culture of *Thermoactinomycetaceae* bacterium *Mechercharimyces asporophorigenens* (YM11–542). Urukthapelstatin A showed inhibition of the growth of human lung cancer A549 cells (IC_{50} = 12 nM) and also showed potent cytotoxic activity against other human cancer cell lines.

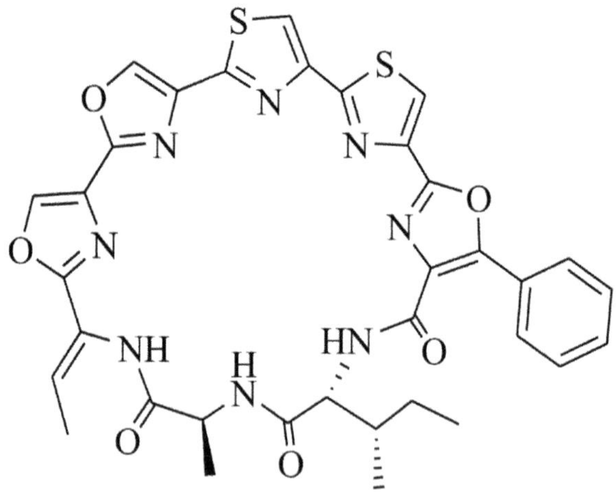

FIGURE 5.8   Urukthapelstatin A.

Two cyclic hexapeptides, Venturamide A and B (Fig. 5.9) have been iso-lated from the marine cyanobacterium *Oscillatoria* species. Venturamide

A and B showed in vitro antimalarial activity against *Plasmodium falciparum* and mild cytotoxic activity against mammalian Vero cells (IC_{50} values of 86 and 56 μM, respectively). They also exhibited mild activity against *Trypanasoma cruzi* and *Leishmania donovani*.

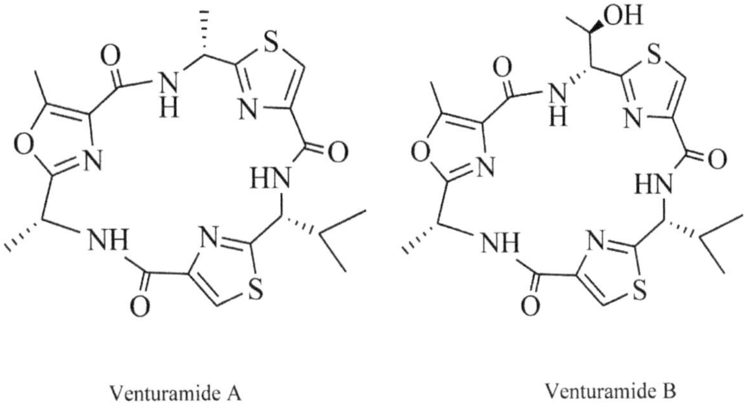

Venturamide A   Venturamide B

FIGURE 5.9   Venturamides A and B.

Four hexacyclopeptides, Aerucyclamide A, B, C and D (Fig. 5.10) have been reported from the toxic freshwater cyanobacterium *Microcystis aeruginosa* (Portmann et al., 2008). Thise compounds contain oxazoline, thiazoline, thiazole or oxazole heterocycles in their structures. Aerucyclamide B showed activity against the chloroquine-resistant strain of *P. falciparum* at sub micro molar concentration (IC_{50} = 0.7 μM). In addition, this compound exhibited a large selectivity for the parasite. In contrast, Aerucyclamides A, C and D displayed low activity against *P. falciparum* at micro molar concentration. Aerucyclamides C also exhibited activity against *T. brucei rhodesiense*.

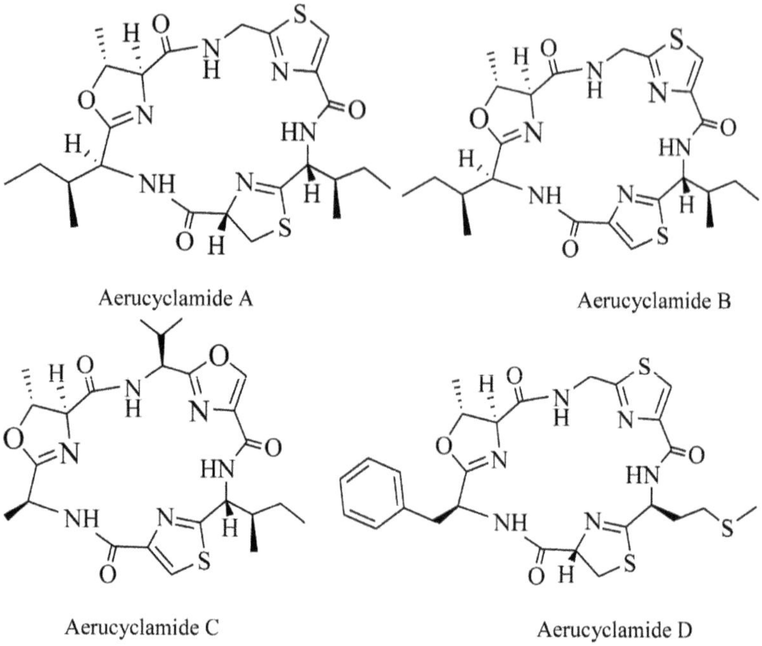

Aerucyclamide A

Aerucyclamide B

Aerucyclamide C

Aerucyclamide D

FIGURE 5.10   Aerucyclamides A–D.

Hexamollamide (Fig. 5.11) a novel hexapeptide, has been isolated from an Okinawan ascidian *Didemnum molle* which showed moderate cytotoxicity against HeLa S3 cells (IC_{50} = 17 µg/mL).

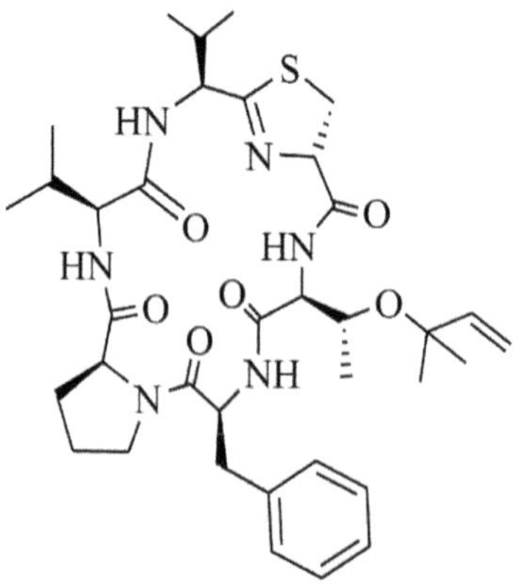

FIGURE 5.11   Hexamollamide.

Two cyclic hexapeptides, Mollamides B and C (Fig. 5.12) have been isolated from *Didemnum molle* (Donia et al., 2008). Mollamide B showed promising antimalarial activity against *Plasmodium falciparum and* also exhibited activity against *Leishmania donovani.* It also showed cytotoxicity against several cell lines. But Mollamide C did not show any activity.

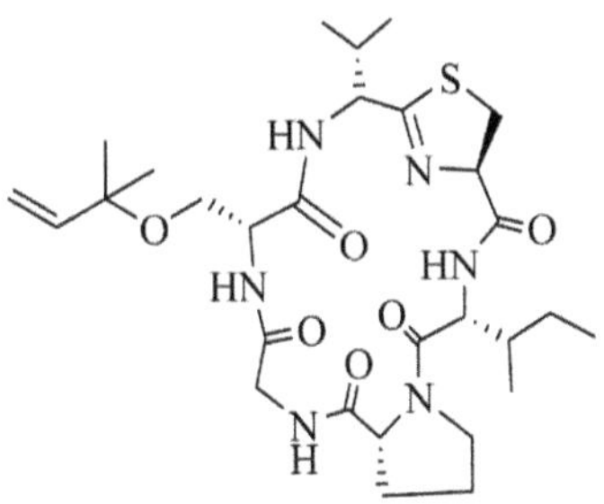

Mollamide B 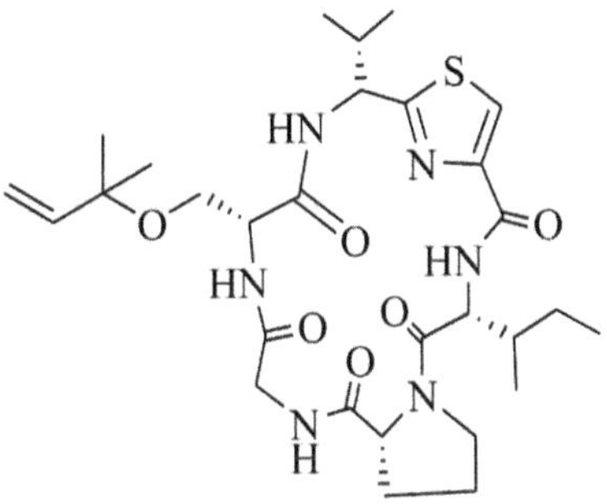   Mollamide C

FIGURE 5.12   Mollamides B and C.

Recently, antiproliferative agent largazole (Fig. 5.13) was isolated from a cyanobacterium of the genus *Symploca* (Taori et al., 2008), which showed promising inhibitory activity of the growth of highly invasive transformed human mammary epithelial cells (MDA–MB–231).

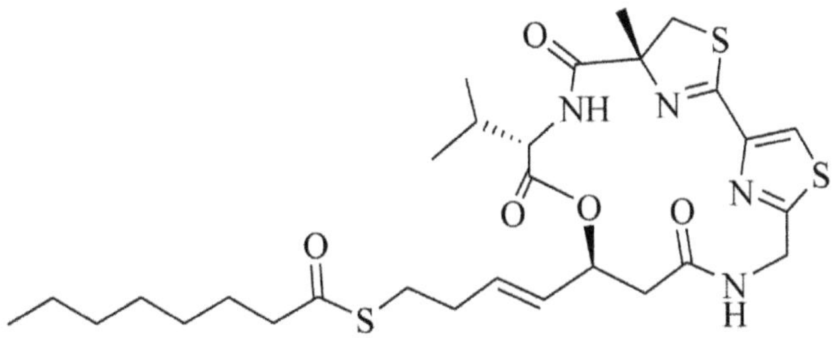

FIGURE 5.13   Largazole.

5.3   SYNTHESIS AND BIOLOGICAL ACTIVITIES OF SOME THIAZOLE DERIVATIVES

In the recent literature, thiazole ring as pharmacophore have been investigated for various pharmacological activities. The studies confirmed that thiazole derivatives (as bioisoster of the imidazole ring) are good pharmacophores for the design of bioactive molecules. In many cases they have shown better therapeutic and less adverse effects as compared to existing class of drugs. They emerged as compounds showing a different mode of action, inhibiting the microbial growth.

Various researchers group have synthesized thiazole derivatives and evaluated their pharmacological activities to find better chemotherapeutic agents. For example, Karegoudar. [23] have synthesized several thiazole derivatives containing 2,3,5-trichlorophenyl moiety (Fig. 5.14) and studied their antibacterial and antifungal activities. In particular, the thiazoles carrying 4-(methylthio) phenyl, salicylamide, N-methylpiparazino and 4,6-dimethyl–2-mercaptopyrimidine substituents, exhibited the highest antibacterial activity. On the other hand, thiazoles carrying 3-pyridyl, biphenyl and 4-mercaptopyrazolopyrimidine substituents exhibited the highest antifungal activity.

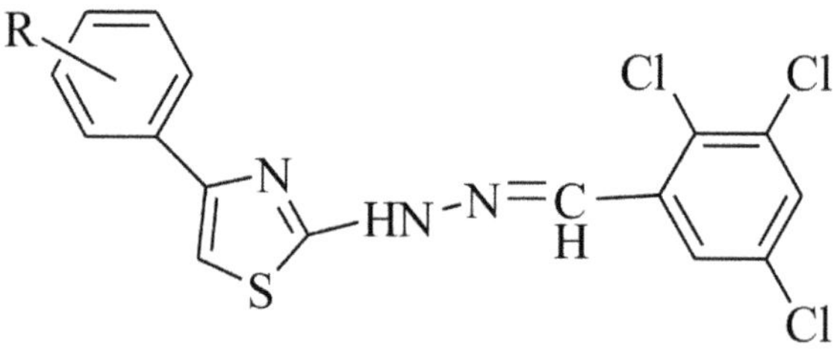

FIGURE 5.14   Thiazole derivatives containing 2,3,5-trichlorophenyl moiety.

Similarly, Ref. [20] have synthesized 2-(arylidenehydrazino)–4–(2,4-dichloro-5-fluorophenyl) thiazoles (Fig. 5.15) and studied their antibacterial and anti-inflammatory activities. Some of the compounds showed good antibacterial and some showed excellent anti-inflammatory activities. Particularly, thiazoles carrying 4-chloro and 4-methoxy substituents showed excellent antibacterial activity against S. aureus and P. aeruginosa. 4-bromophenyl, 4-methylphenyl and 4-chloro substituents showed acute anti-inflammatory activity. Compounds carrying 4-bromophenyl and 3,4-methylenedioxybenzylidene amino substituents showed excellent chronic anti-inflammatory activity comparable with that of Ibuprofen.

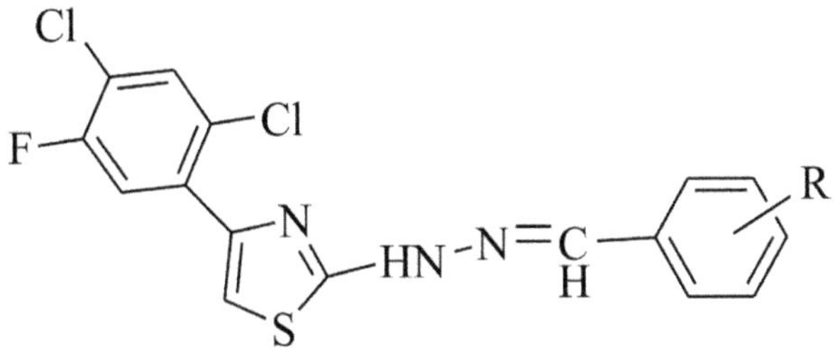

R = H, 4-Cl, 2,4-Cl$_2$, 4-OCH$_3$

FIGURE 5.15   2-(arylidenehydrazino)-4-(2,4-dichloro-5-fluorophenyl) thiazoles.

Ding [13] synthesized several thiazolone based sulfonamides (Fig. 5.16) using various hetero-aryl sulfonyl chlorides and different aldehydes, and evaluated for their inhibitory activity of NS5B polymerase, to target HCV. Some of them showed inhibitory activity, IC$_{50}$ value at 0.6 nM concentration.

Crute [10] reported thiazole derivatives (Fig. 5.16) exhibiting high antiviral activity against hepatitis C virus (HCV) and HSV.

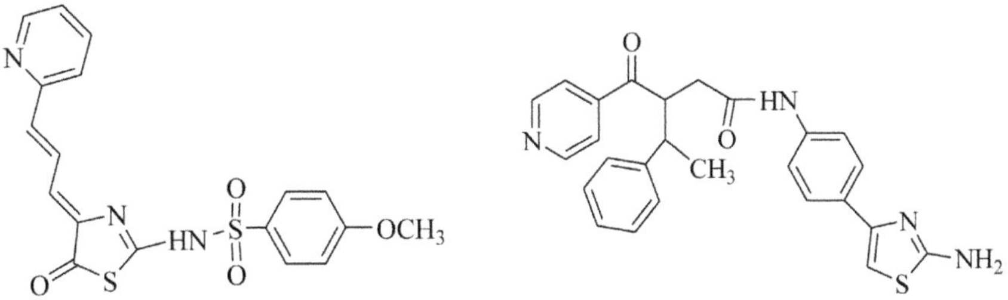

FIGURE 5.16   Thiazole derivatives showing antiviral activities.

El-Sabbagh et al. (2009) have synthesized new pyrazole and thiazole derivatives (Fig. 5.17) and evaluated for antiviral activity against a broad panel of viruses in different cell cultures. They reported that N-acetyl 4,5-dihydropyrazole showed activity at subtoxic concentrations against vaccinia virus (Lederle strain) in HEL cell cultures with a 50% effective concentration (EC$_{50}$) value of 7 μg/mL.

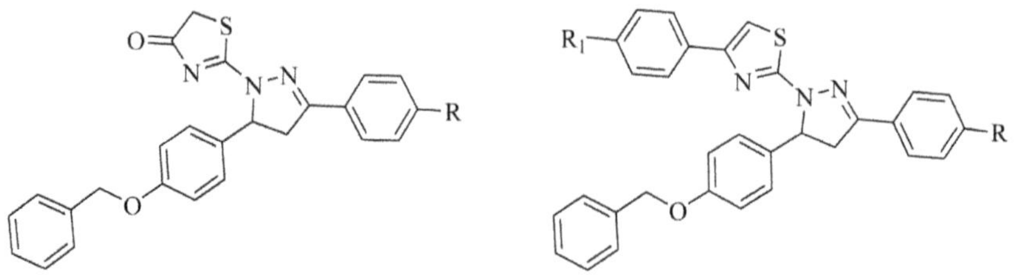

FIGURE 5.17   Thiazole derivatives showing antiviral and anticancer activities.

Bondock. [3] synthesized some new thiazole, thiazolidinone and thia-zoline derivatives starting from 1-chloro-3,4-dihydronaphthalene-2-car-boxaldehyde and evaluated their antimicrobial activities (Fig. 5.18). Thise compounds were screened in vitro for their antimicrobial activities against three strains of bacteria *Bacillus subtilis, Bacillus megaterium, Escherichia coli,* and two strains of fungi *Aspergillus niger and Aspergillus oryzae* by the agar diffusion technique. The results showed that most of the thiazolidinones showed comparable activity, the thiazole derivatives, thiazoline derivatives and thiazolo [5, 4–d] pyrimidine derivatives showed very high activity with respect to the used references. Nearly all of the compounds exhibited high antifungal activity against the reference drug.

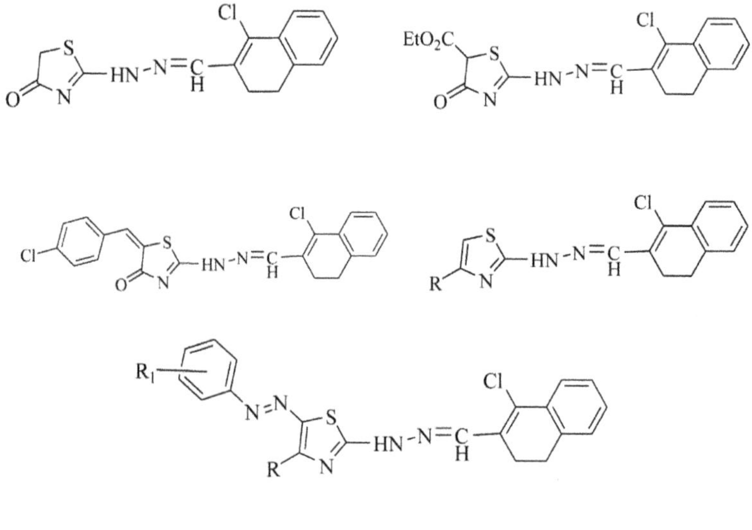

R= Me, Phe; R₁= 4-Me, 4-Cl

FIGURE 5.18   Thiazolo-1-chloro-3,4–dihydronaphthalene derivatives.

Bekhit. [5] synthesized a series of pyrazolyl benzenesulfonamide derivatives (Fig. 5.19) by cyclization of the intermediate N,N-dimethylaminomethylene-4[3-phenyl-4-(substituted thiosemicarbamoyl hydrazonomethyl)-1H-pyrazol-1-yl]benzenesulfonamide with ethyl bromoacetate to afford the corresponding thiazolidinyl derivatives. All the compounds showed anti-inflammatory activity and three of them surpassed that of indomethacin both locally and systemically in the cotton pellet granuloma and rat paw edema bioassay. They reported that the active compounds showed selective inhibitory activity towards COX–2 enzyme as revealed by the in vitro enzymatic assay. All the tested compounds proved to have superior gastrointestinal (GI) safety profiles as compared to indomethacin, when tested for their ulcerogenic effects. The acute toxicity study of compounds having promising anti-inflammatory activity indicated that they are well tolerated both orally and parent rally.

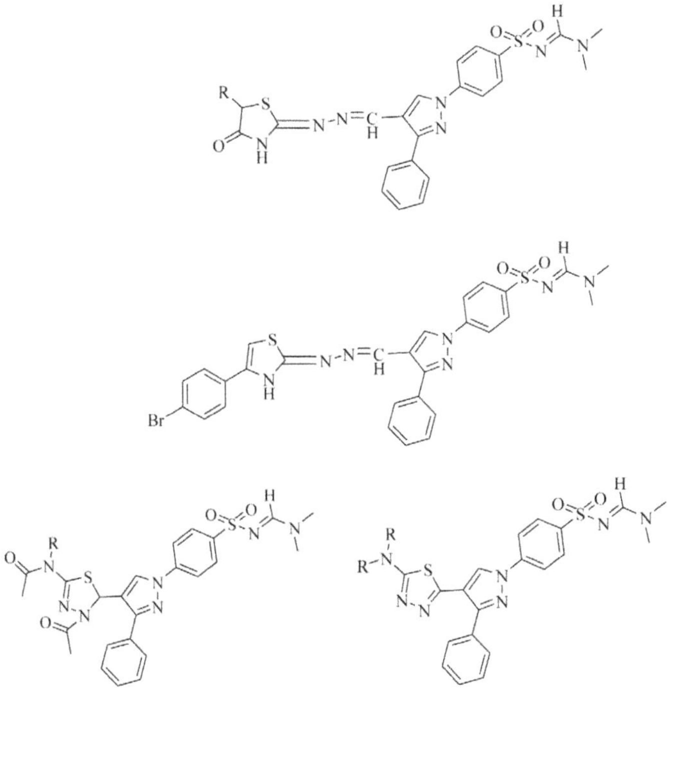

R= C_6H_5, 4-ClC_6H_4, 4-$CH_3C_6H_4$

FIGURE 5.19 Pyrazolyl benzenesulfonamide derivatives.

Antimicrobial activity tests (MIC) of some compounds showed comparable antibacterial activity to that of ampicillin against *Escherichia coli*, while few compounds possessed about half the activity of ampicillin against *Staphylococcus aureus*. The authors also reported that all the tested compounds have weak or no antifungal activity against *Candida albicans*.

Aridoss. [2] synthesized a series of *N*-(*N*-methylpiperazinoacetyl)-2,6-diarylpiperidin-4-ones (Fig. 5.20) by the base catalyzed nucleophilic substitution of *N*-chloroacetyl-2,6-diarylpiperidin-4-ones obtained from their corresponding 2,6-diarylpiperidin-4-ones with *N*-methylpiperazine.

The authors screened the compounds for their possible antibacterial and antifungal activities against a spectrum of microbial agents besides analgesic and antipyretic activities. They found promising results for compound containing R_1 = Me and R = Cl against bacterial and R_1 = Me and R = Me against fungal strains whereas these showed beneficial analgesic and antipyretic profiles at a concentration of 60 mg/kg and were also found to be more potent than the reference drug, ciprofloxacin.

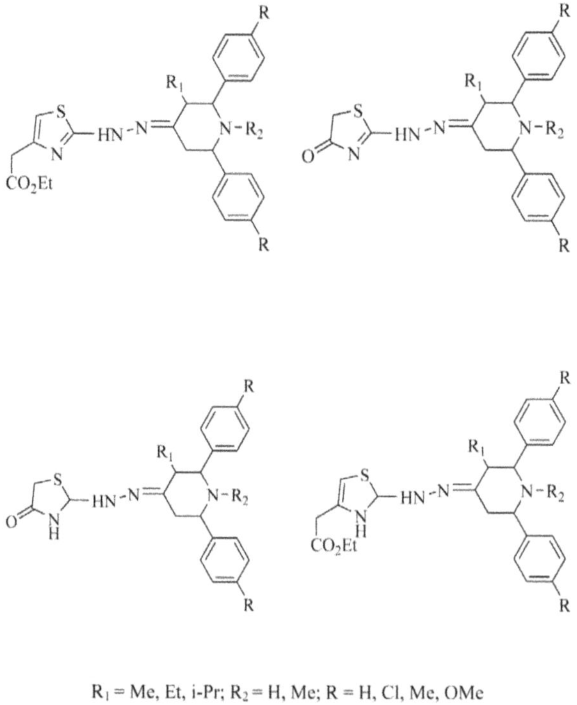

R_1 = Me, Et, i-Pr; R_2 = H, Me; R = H, Cl, Me, OMe

FIGURE 5.20 *N*-(*N*-methylpiperazinoacetyl)-2,6-diarylpiperidin-4-ones.

Bekhit et al. [5] reported the synthesis of two novel series of structurally related 1*H*-pyrazolyl derivatives of thiazole (Fig. 5.21). The synthesized compounds were examined for their in vivo anti-inflammatory activity and in vitro antimicrobial activity against Gram negative and Gram positive bacteria and fungi. Some compounds showed dual anti-inflammatory and antimicrobial activity.

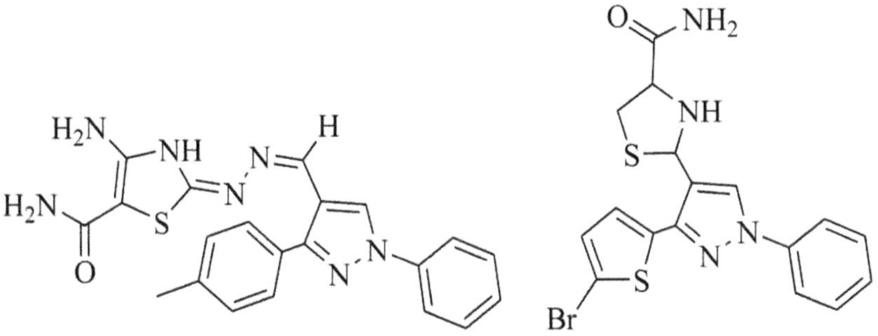

FIGURE 5.21   1*H*-pyrazolyl derivatives of thiazolo [4,5-*d*]pyrimidines.

Turan-Zitouni et al. [36] synthesized *N*-(1-arylethylidene)-*N'*-[4-(indan-5-yl) thiazol-2-yl] hydrazones.

Derivatives (Fig. 5.22) by reacting 1-(1-arylethylidene) thiosemicarbazide with 1-(5-indanyl)-2-bromoethanone. The authors screened the synthesized compounds for their anti tubercular activity which were determined by broth microdilution assay, the Microplate Alamar Blue Assay, in BACTEC12B medium. The results were screened in vitro, using BACTEC 460 Radiometric System against Mycobacterium tuberculosis $H_{37}Rv$ (ATCC 27294) at 6.25 µg/mL and some of the tested compounds showed important inhibition ranging from 92% to 96%. They also investigated for their cytotoxic properties on normal mouse fibroblast (NIH/3T3) cell line and the results showed no significant cytotoxic activity of compounds at the concentrations under 50 µg/mL.

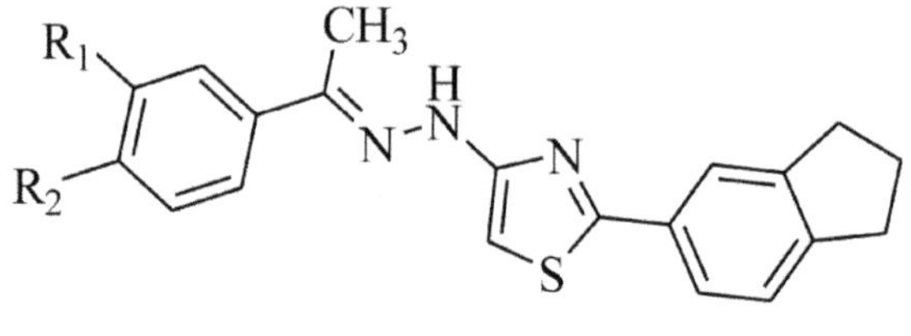

R_1 = H, Cl; R_2 = H, CH_3, OCH_3, NO_2, Cl

FIGURE 5.22   *N*-(1-arylethylidene)-*N'*-[4-(indan-5-yl)thiazol-2-yl]hydrazones
derivatives.

Vijaya Raj et al. (2007) has reported novel 2-bromo-5-methoxy-*N'*-[4-
(aryl)-1,3-thiazol-2-yl]benzohydrazide derivatives (Fig. 5.23) which were
screened for their analgesic, antifungal and antibacterial activities and
three of the compounds were screened for antiproliferative activity. Two
of the newly synthesized compounds exhibited promising analgesic activity and one compound exhibited in vitro antiproliferative activity.
      R= OH, Cl, Br

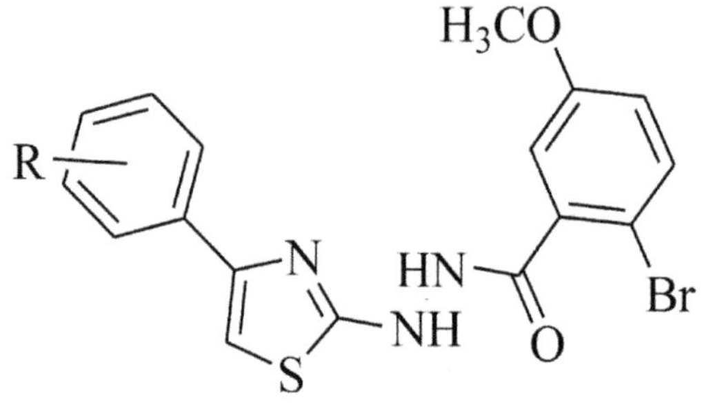

R= OH, Cl, Br

FIGURE 5.23   2-bromo-5-methoxy-*N'*-[4-(aryl)-1,3-thiazol-2-yl]benzohydrazide
derivatives.

Cukurovall [11] have synthesized a series of Schiff bases combining
2,4-disubstituted thiazole and cyclobutane rings, and hydrazone moieties

in the same molecule (Fig. 5.24). The authors evaluated thise compounds for antibacterial and antifungal activities on microorganisms including four bacteria and fungus *Candida tropicalis*. They reported both the antibacterial and antifungal activities and MIC values of the compounds. Among the tested compounds, the most effective compound providing a MIC value of 16 µg/mL are compound having R_2 = OH against *C. tropicalis* and *Bacillus subtilis* and compound having R_3 = Br against *B. subtilis*.

Ar = Ph, p-Xylyl, Mesityl; R_1 = H, OCH_3, Cl, Br; R_2 = OH, OCH_3; R_3 = OCH_3, NO_2, Cl, Br

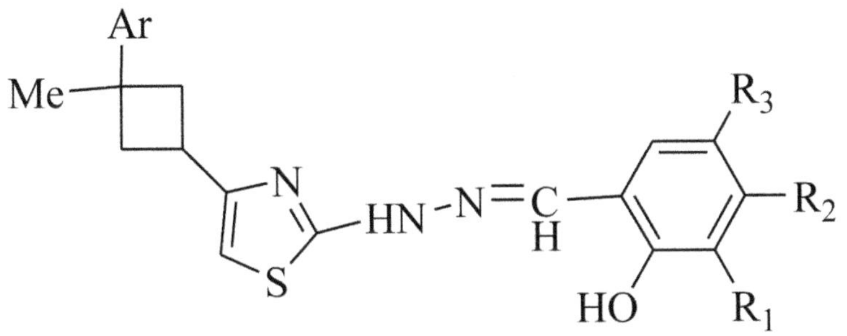

FIGURE 5.24   Schiff bases containing 2,4-disubstituted thiazole ring.

KEYWORDS

- antimicrobial agents
- biological activities
- Chemistry of Thiazole
- Pharmacokinetic properties
- Scaffold
- Therapeutic

REFERENCES

1. Aridoss, G., Parthiban, P., Ramachandran, R., Prakash, M., Kabilan, S., & Jeong, Y. T. (2009). Synthesis and spectral characterization of a new class of N-(N-methylpi-

perazinoacetyl)-2, 6-diarylpiperidin-4-ones: Antimicrobial, analgesic and antipyretic studies, *European Journal of Medicinal Chemistry*, 44, 577–592.

2. Aridoss, G., Amirthaganesan, S., Kim, M. S., Kim, J. T., & Jeong, Y. T. (2009). Synthesis, spectral and biological evaluation of some new thiazolidinones and thiazoles based on t-3-alkyl-r-2, c-6-diarylpiperidin-4-ones, *European Journal of Medicinal Chemistry*, 44, 1–12.

3. Bell, F. W., Cantrell, A. S., Hoegberg, M., Jaskunas, S. R., Johansson, N. G., Jordan, C. L., Kinnick, M. D., Lind, P., & Jr. Morin, J. M. (1995). Phenethylthiazolethiourea, PETT Compounds, a New Class of HIV-1 Reverse Transcriptase Inhibitors. 1. Synthesis and Basic Structure-Activity Relationship Studies of PETT Analogs, *J. Med. Chem.*, 38, 4929–36.

4. Bekhit, A. A., Ashour, H. M. A., Ghany, Y. S. A., El-Din, A., Bekhit, A., & Baraka, A. (2008). Synthesis and biological evaluation of some thiazolyl and thiadiazolyl derivatives of 1H-pyrazole as anti-inflammatory antimicrobial agents, *European Journal of Medicinal Chemistry*, 43, 456–463.

5. Bekhit, A. A., Fahmy, H. T. Y., Rostom, S. A. F., & Baraka, A. M. (2003). Design and synthesis of some substituted 1*H*-pyrazolyl-thiazolo [4, 5-*d*] pyrimidines as anti-inflammatory–antimicrobial agents, Eur. J. Med. Chem., 38, 27–36.

6. Bondock, S., Khalifa, W., & Fadda, A. A. (2007). Synthesis and antimicrobial evaluation of some new thiazole, thiazolidinone and thiazoline derivatives starting from 1-chloro-3,4-dihydronaphthalene-2-carboxaldehyde, *European Journal of Medicinal Chemistry*, 42, 948–954.

7. Borisenko, V. E., Koll, A., Kolmakov, E. E., & Rjasnyi, A. G. (2006). Hydrogen bonds of 2-aminothiazoles in intermolecular complexes 1:1 and 1:2 with proton acceptors in solutions *J. Mol. Struct* 783, 101–115.

8. Breslow, R. (1958). On the Mechanism of Thiamine Action IV.1 Evidence from Studies on Model Systems, *J. Am. Chem. Soc.* 80, 37, 19–26.

9. Buchanan, J. L., Bohacek, R. S., Luke, G. P., Hatada, M., Lu, X., Dalgarno, D. C., Narula, S. S., Yuan, R., & Holt, D. A. (1999). Structure-based design and synthesis of a novel class of Src SH2 inhibitors, *Bioorg. Med. Chem. Letts.*, 9, 2353–2358.

10. Crute, J. J., Grygon, C. A., Hargrave, K. D., Simoneau, B., Faucher, A. M., Bolger, G., Kiber, P., Liuzzi, M., & Cordingley, M. G. (2002). Herpes simplex virus helicase-primase inhibitors are active in animal models of human disease *Nat. Med.*, 8, 386–391.

11. Cukurovali, A., Yilmaz, I., Gur, S., & Kazaz, C. (2006). Synthesis, antibacterial and antifungal activity of some new thiazolyl hydrazone derivatives containing 3-substituted cyclobutane ring, *Eur. J. Med. Chem.*, 41, 201–207.

12. De Souza, M. V. N., De, & Almeida, M. V. (2003). Drug antiHIV: Past, present and future perspectives, *Quim. Nova* 26, 366–372.

13. Ding, Y., Smith, K. L., Varaprasad, C. V. N. S., Chang, E., Alexander, J., & Yao, N. (2007). Synthesis of thiazolone-based sulfonamides as inhibitors of HCV NS 5B polymerase, °*Bioorg. Med. Chem. Lett.* 17, 841–845.

14. El Kazzouli, S., Berteina-Raboin, S., Mouaddib, A., Guillaumet, G. (2002). Solid support synthesis of 2,4-disubstituted thiazoles and aminothiazoles, *Tetrahedron Letters*, 43, 3193–3196.

15. El-Subbagh, H. I., Al-Obaid, A. M. (1996). 2,4-disubstitued thiazoles II. A novel class of antitumor agents, synthesis and biological evaluation, *Eur. J. Med. Chem.* 31, 1017–1021.

16. Fink, B. E., Mortensen, D. S., Stauffer, S. R., Aron, Z. D., & Katzenellenbogen, J. A. (1999). Novel structural templates for estrogen-receptor ligands and prospects for combinatorial synthesis of estrogens, *Chem Biol.* 6, 205–219.

17. Hallinan, E. A., Hagen, T. J., Tsymbalov, S., Stapelfeld, A., & Savage, M. A. (2001). 2,4-Disubstituted Oxazoles and Thiazoles as Latent Pharmacophores for Diacylhydrazine of SC-51089, a Potent PGE_2 Antagonist, *Bioor. Med. Chem.* 9, 1–6.

18. Hantzsch, A., & Weber, J. H. (1887). *Ber. Dtsch. Chem. Ges.* 20, 3118–32

19. Haviv, F., Ratajczyk, J. D., DeNet, R. W., Kerdesky, F. A., Walters, R. L., Schmidt, S. P., Holms, J. H., Young, P. R., & Carter, G. W. (1988). 3-[1–2- Benzoxazolyl hydrazine propanenitrile] derivatives: inhibitors of immune complex induced inflammation, *J. Med. Chem.* 31, 1719–1728.

20. Holla, B. S., Malini, K. V., Rao, B. S., Sarojini, B. K., & Kumari, N. S. (2003). Synthesis of some new 2,4-disubstituted thiazoles as possible antimicrobial and antiinflammatory agents, *Eur. J. Med. Chem.*, 38, 313–318.

21. Kabalka, G. W., & Mereddy, A. R. (2006). Microwave promoted synthesis of functionalized 2-aminothiazoles, *Tetrahedron Lett*, 47, 5171–5172.

22. Karegoudar et al., (2008). Synthesis of some novel 2,4-disubstituted Thiazoles as possible antimicrobial agents, *Eur. J. Med. Chem.*

23. Karegoudar, P., Karthikeyan, M. S., Prasad, D. J., Mahalinga, M., Holla, B. S., & Kumari, N. S. (2008). Synthesis of some novel 2,4-disubstituted thiazoles as possible antimicrobial agents, *Eur. J. Med. Chem.* 43, 261–267.

24. Knadler, M. P., Bergstrom, R. F., Callaghan, J. T., & Rubin, A. (1986). Nizatidine, an H_2-blocker. Its metabolism and disposition in man, *Drug Metab. Dispos.*, 14, 175–182.

25. Kolb, J., Beck, B., Almstetter, M., Heck, S., Herdtweck, E., & Domling, A. (2003). New MCRs: The first 4-component reaction leading to 2, 4-disubstituted thiazoles, *Molecular Diversity* 6, 297–313.

26. Kumar, Y., Green, R., Borysko, K. Z., Wise, D. S., Wotring, L. L., & Townsend, L. B. (1993). Synthesis of 2,4-disubstituted thiazoles and selenazoles as potential antitumor and antifilarial agents: 1. Methyl 4-,Isothiocyanatomethyl thiazole-2-carbamates,-selenazole-2-carbamates, and related derivatives, *J. Med. Chem.* 36, 3843–3848.

27. Kumar, Y., Green, R., Wise, D. S., Wotring, L. L., & Townsend, L. B. (1993).Synthesis of 2,4-disubstituted thiazoles and selenazoles as potential antifilarial and antitumor agents. 2. 2-arylamido and 2-alkylamido derivatives of 2-amino-4-, isothiocyanatomethylthiazole and 2-amino-4-, isothio cyanatomethylselenazole, *J. Med. Chem.*, 36, 3849–3852.

28. Lednicer, D., Mitscher, L. A., & George, G. I. (1990). Organic Chemistry of Drug Synthesis, *Wiley New York*, 4, 95–97.

29. Medime, E. & Capan, G. (1994). Synthesis and anticonvulsant activity of new 4-thiazolidone and 4- thiazoline derivatives. *Farmaco* 49, 449–451.

30. Metzger, J. V., Katritzky, R., & Rees, C. W. (1984). Eds. Pergamon, Comprehensive Heterocyclic Chemistry, NY, 6, 235–332.

31. Milne, G. W., & Ashgate, A. Ed., (2000). Handbook of Antineoplastic Agents, Gower, London, UK.

32. Patt, W. C., Hamilton, H. W., Taylor, M. D., Ryan, M. J., Taylor, Jr D. G., Connolly, C. J., Doherty, A. M., Klutchko, S. R., & Sircar, I. (1992). Structure-activity relationships of a series of 2-amino-4-thiazole-containing renin inhibitors, *J. Med. Chem.*, 35, 2562–2572.

33. Revol-Junelles, A. M., Mathis, R., Krier, F., Delfour, A., & Lefebvre, G. (1996). Leuconostoc mesenteroides subsp Mesenteroides FR52 synthesize two distinct bacteriocins, *Lett. Appl. Microbiol*, 23, 120–124

34. Siddiqui, H. L., Iqbal, A., Ahmed, S., & Weaver, W. (2006). Synthesis and Spectroscopic Studies of New Schiff Bases, *Molecules*, 11, 206–211.

35. Tsuji, K., & Ishikawa, H. (1994). Synthesis and antipseudomonal activity of new 2-isocephems with a dihydroxypyridone moiety at C–7, *Bioorg. Med. Chem. Lett.*, 4, 1601–1606.

36. Turan-Zitouni, G., zdemir, A., Kaplancikli, Z. A., Benkli, K., Chevallet, P., & Akalin, G. (2008). Synthesis and antituberculosis activity of new thiazolylhydrazone derivatives, *European Journal of Medicinal Chemistry*, 43, 981–985.

37. Van Muijlwijk-Koezen, J. E., Timmerman, H., Vollinga, R. C., Von Drabbe Kunzel, J. F., De Groote, M., Visser, S., & IJzerman, A. P. (2001). Thiazole and Thiadiazole Analogues as a Novel Class of Adenosine Receptor Antagonists, *J. Med. Chem.* 44, 749–762.

38. Wang, M., Wang, L. F., Li, Y. Z., Li, Q. X., Xu, Z. D., & Qu, D. M. (2001). Antitumor activity of transition metal complexes with the thiosemicarbazone derived from 3-acetylumbelliferone, *Trans. Met. Chem.* 26, 307–310.

39. Wilson, K. J., Illig, C. R., Subasinghe, N., Hoffman, J. B., Rudolph, M. J., Soll, R., Molloy, C. J., Bone, R., Green, D., Randall, T., Zhang, M., Lewandowski, F. A., Zhou, Z., Sharp, C., Maguire, D., Grasberger, B., DesJarlais, R. L., & Spurlino, J. (2001). Synthesis of thiophene-2-carboxamidines containing 2-amino-thiazoles and their biological evaluation as urokinase inhibitors. *Bio. Org & Med. Chem. Lett*, 11, 915–918.

40. Yu, H., & Adedoyin, A. (2003). ADME-Tox in drug discovery: Integration of experimental and computational technologies, *Drug Discov. Today*, 8, 852.

CHAPTER 6

A NOTE ON NEW BIOLOGICALLY ACTIVE COMPOSITE MATERIALS ON THE BASIS OF DIALDEHYDE CELLULOSE

AZAMAT A. KHASHIROV, AZAMAT A. ZHANSITOV, GENADIY E. ZAIKOV, and SVETLANA YU. KHASHIROVA

CONTENTS

ABSTRACT

In this work for the first time have been studied modification peculiarities of microcrystalline cellulose (MCC) and its oxidized form (dialdehyde cellulose DAC) guanidine-containing monomers and polymers of vinyl and diallyl series. Researched the structure of the composites by IR spectroscopy and SEM. The biological activity of the synthesized composite materials was investigated and shown that the composite synthesized materials are quite active and have a biocidal effect against Gram-positive (St. Aureus) and Gram (E. coli) microorganisms.

6.1 INTRODUCTION

Formation and research of systems "polymeric carrier" biologically active substance have lately got great importance. Such systems find an application as immobilized biocatalysts, bioregulators and an active form of medicinal substances of the prolonged action [1].

In this work for the first time have been studied modification peculiarities of MCC and its oxidized form (DAC) guanidine-containing monomers and polymers of vinyl and diallyl series. The structure and some characteristics of used guanidine-containing modifiers are shown in Table 6.1.

TABLE 6.1 Structure and Some Characteristics of Guanidine-Containing Modifiers of Cellulose and Dialdehyde Cellulose

Modifier	Molecular Weight	Melting Point, °C	Structure
Acrylate guanidine (AG)	131,134	175–176	R=H
Methacrylate guanidine (MAG)	145,160	161–163	R=CH3

| N,N-Diallylguanidine acetate (DAGA) | 199,253 | 211–212 | 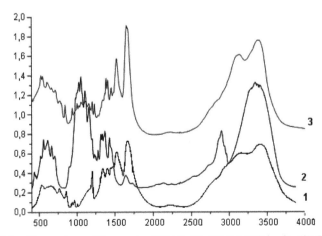 |
| N,N-Diallylguanidine trifluoroacetate (DAGTFA) | 253,224 | 157–158 | |

6.2 RESULTS AND DISCUSSION

Composite materials were received by treating the MCC or DAC, water-soluble monomeric guanidine derivatives (Table 6.1), with subsequent polymerization. Quantity of the monomer/polymer twitters ionic quaternary ammonium cations acrylate and diallyl guanidine derivatives included in MCC or DAC determined by nitrogen content using elemental analysis.

The results of IR spectroscopic studies show the structural differences of cellulose samples and its modified forms (Fig. 6.1).

FIGURE 6.1 Comparison of the spectra PMAG (1), DAC-MAG in situ (2), DAC (3).

For example, the polymerization in DAC MAG in situ (Fig. 6.1, curve 3) in the spectra varies the ratio of the intensity of stripes both cellulose (within 1000–1100 cm⁻¹) and MAG, moreover, the stripe disappears within 860 cm⁻¹ indicating the presence of a double tie. Splitting of the tie C=O ties PMAG within 1250 cm⁻¹ is taking place that clearly shows strong mutual influence of the DAC and MAG/PMAG and formation of bimatrix systems. Increasing the intensity of the peak 1660 cm⁻¹ in the spectrum of the DAC–MAG formation aldimine connection. Increasing the width of the characteristic absorption stripes in DAC–MAG within 1450–1680 cm⁻¹, probably due to the formation of relatively strong ties with the active MAG DAC centers.

Comparison of the IR spectra of the composites and DAZ DAZ–DAG and DAZ–DAGTFA demonstrates differences in the spectra of thise compounds and indicates the formation of new structures (Fig. 6.2).

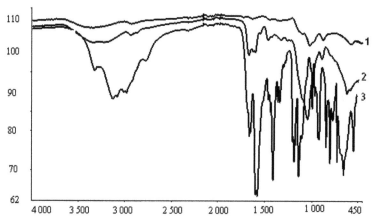

FIGURE 6.2   IR-spectra DAC (1), DAC-DAGTFA (2); DAGTFA (3).

In the spectra of the modified DAC v_{OH} observed increase in absorption from the high-frequency, especially in the spectrum of DAC modified DAG. This is due to increasing of the hydroxyls involved in weak hydrogen bonds. Stretching vibrations of C–H bonds of methane and methylene groups DAC appear in 3000–2800 cm⁻¹. In the spectra of the modified

DAG and DAGTFA DAC these valence oscillations are superimposed on the absorption of the CH$_2$ groups that are part of diallyl compounds.

This leads to an increase in the intensity of the absorption bands with a frequency of ~2900 cm^{-1}. In the area of ~1650 cm^{-1} peaks appear adsorbed water. Raising polar amino group comprising the modifier that increases the polarity of the substrate that assists in keeping the surface modified pulp samples more water adsorption due to hydrogen bonds. Increasing the intensity of the peak 1655 cm^{-1} in the spectrum of DAC–DAG and DAC–DAGTFA may indicate formation aldimine communication giving a signal in this area. In the polymerization in the DAG and DAGTFA, DAC in situ peak at 1140 cm^{-1} present in the DAC disappears. Obviously, the terminal CHO groups DAC and guanidine containing diallyl modifier reacted with each other.

Thus, immobilization AG, MAG, DAG and DAGTFA, DAC between components in the formation of various types of bonds: due to van der Waals forces, intra and intermolecular coordination and hydrogen bonds, C–C bonds formed during the free radical polymerization in situ immobilized AG, MAG, DAG and DAGTFA, bonds formed during the graft copolymerization of monomer radical salts with DAC and labile covalent aldimine C=N bonds formed by reacting aldehyde groups with amino DAC guanidine containing compounds.

Composite materials obtained by polymerization and DAG, DAGTFA in situ in the inter and intra fibrillar DAC pores, dissolve well in water. We can assume that the action of pulp and DAGTFA, DAG, a major role in breaking the inter and intra molecular hydrogen bonds play anions CH$_3$COO$^-$ and SF$_3$COO$^-$ which form a DAC stable complexes through its hydroxyl groups. Simultaneously, the esterification may occur partly sterically more accessible alcohol groups DAC. SEM method shows that the dissolution of cellulose sphere-like type of complex is formed between the components of the solution in which the cellulose macromolecules have coil conformation (Fig. 6.3).

In the process of dissolution of DAC and DAG, DAGTFA also act as acceptors of hydrogen bonds and associated solvent molecules, thereby preventing reassociation of cellulose macromolecules.

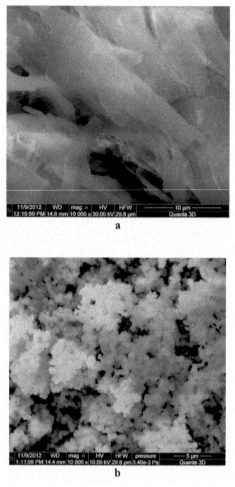

FIGURE 6.3   SEM images dialdehyde cellulose (a) and complex of dialdehyde cellulose with N,N-diallylguanidine trifluoroacetate (b)

The biological activity of the synthesized composite materials was investigated and shown that the composite synthesized materials are quite active and have a biocidal effect against Gram-positive (St. Aureus) and Gram (E. coli) microorganisms. Being the most expressed biocidal activity is shown in the composites with diallyl derivatives of guanidine.

6.3 CONCLUSION

Studies of the structure of the composites by SEM showed that DAG and DAGTFA localized in the surface layers of the composite, which increases the availability of biocidal centers and explains a higher relative activity of these composites.

In the case of dialdehyde cellulose, modified acrylate derivatives, guanidine antibacterial active groups are in the deeper layers of the interfibrillar dialdehyde cellulose, which reduces their bioavailability and therefore, the DAC-PAG and DAC-PMAG and start to show the bactericidal activity of only 48 h. Slowing down the release rate of the bactericidal agent opens prospect of long-acting drugs with a controlled-release bactericide.

KEYWORDS

- Acrylate guanidine
- Cellulose
- IR spectroscopy
- Methacrylate guanidine
- N,N-diallylguanidine acetate
- N,N-diallylguanidine trifluoroacetate
- SEM

REFERENCES

1. Mikitaev, A. K., Kozlov, G. V., & Zaikov, G. E. (2009). *Polymer Nanocomposites: A variety of structural forms and applications*, Moscow: Science, 278 p.

CHAPTER 7

A TECHNICAL NOTE ON NEW NANOCOMPOSITES BASED ON LAYERED ALUMINOSILICATE AND GUANIDINE CONTAINING POLYELECTROLYTES

AZAMAT A. KHASHIROV, AZAMAT A. ZHANSITOV,
GENADIY E. ZAIKOV, and SVETLANA YU. KHASHIROVA

CONTENTS

ABSTRACT

The new functional nanomaterials based on layered alumino silicate and guanidine containing polyelectrolytes combining high bactericidal activity with an increased ability to bind to heavy metals and organic pollutants were received. To prove the chemical structure of the model compounds (zwitterionic delocalized resonance structures AG/MAG and PAG/ PMAG), as well as the presence of such structures in nanocomposites received on their basis and the MMT, IR, ^{1}H NMR spectroscopy, X-ray diffraction studies and nanoindentation/sclerometry followed by scanning the surface in the area of the indentation were used.

7.1   INTRODUCTION

One of the perspective ways of receiving functional nanocomposites is the intercalation of ionic monomers into inorganic layered materials such as clay minerals and polymerization of the last ones in situ. Such approach is of versatile interest. First, practical opportunity is provided for production of layered nano composites. Secondly, it is important for its unusual intercalation physicochemistry and its manifestation in the acquisition of improved physicochemical properties systems. Furthermore, the study of these products can give important information about the chemical nature of these interactions, the specificity of adsorption of ionic monomers and polymers of nanoscale particles, etc.

From the literature it is known that the preparation of organo modified nanostructures based on natural montmorillonite clays and organic monomers and polymers having difficulties because of their incompatibility [1]. For hydrophobizing the basal surfaces of the calcium (Ca^{2+}-MMT) and sodium (Na^+-MMT) montmorillonite in this study was first used, not only the monomers of the vinyl series containing cation-tropic quaternary imine derivative salt of acrylate/methacrylate-guanidine (AG/MAG) but the polymers (PAG/PMAG) based on them, obtained in situ.

7.2   EXPERIMENTAL PART

Polymerization of monomers containing guanidine performed in situ in the presence of ammonium persulfate (PSA). The optimal polymerization conditions were determined (temperature 60 °C, the mass of PSA was 1% from the weight of the monomer, time 60 min). In this article the main accent was done on the hybrid nanocomposites matrix of which was flaky alumino silicate montmorillonite (MMT), with various degrees of filling (Ca^{2+}) Na^+-MMT: MAG and (Ca^{2+}) Na^+-MMT:PMAG.

To prove the chemical structure of the model compounds (zwitterionic delocalized resonance structures AG/MAG and PAG/PMAG), as well as the presence of such structures in nanocomposites received on their basis and the MMT, IR, ^1H NMR spectroscopy, X-ray diffraction studies and nanoindentation/sclerometry followed by scanning the surface in the area of the indentation were used.

IR spectral studies of acrylate and methacrylate guanidine adsorbed Na– and Ca– montmorillonite showed differences due to the influence of exchangeable cations on the strength and mechanism of the molecules AG and MAG with mineral surface. Thus, in the interpacket space of sodium montmorillonite hypertension and MAG are not formed due to surface oxygen atoms. However, it is shown that in the case of Ca–montmorillonite molecule containing guanidine salts enter into specific interactions with exchangeable cations of the mineral and simultaneously form hydrogen ties with surface oxygen atoms, or the neighboring atoms of the adsorbate. Favorable condition for the participation of amino groups AG and MAG in simultaneous interaction with exchangeable cations and surface oxygen atoms also creates a flat orientation of the molecules of acrylate and methacrylate guanidine in interlayer galleries mineral that has been shown by scanning probe microscopy.

This method confirmed the formation of the polymerization reaction of MAG in situ polymeric quaternary ammonium cations in nanostructures, macromolecules of which are arranged in ordered parallel interlayer galleries (Fig. 7.1a, b). Their interaction with active functional centers montmorillonite basal surfaces Si–O–Si–OH and the results indicate the indentation. So, if one of the faces of a three-sided pyramid Berkovich (indentation) is parallel to the interlayer galleries Ca^{2+}–MMT, the indentation selection PMAG ("bulk") is the smallest (Fig. 7.1b). To the existence of such interactions indirectly indicates an increase in the amount released from the interlayer galleries PMAG (height "bulk" faces in (Fig. 7.1) with increasing load on the indenter.

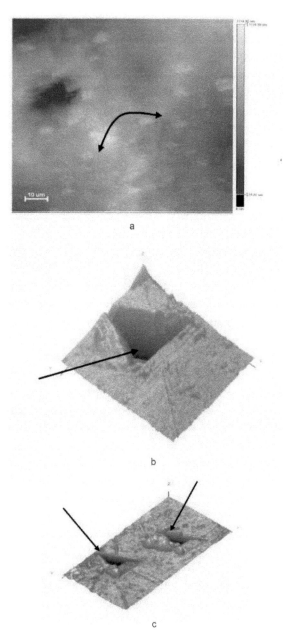

FIGURE 7.1 (a) Topography image of the surface of the composite Na⁺-MMT:PMAG (50:50), (b) three-dimensional image of the relief (indent), and c) three-dimensional image of the relief (indent at different load).

X-ray diffraction studies showed the formation of intercalated and ex-foliated nanostructures by radical polymerization of the monomer AG and MAG in situ.

Nanocomposites were prepared with a degree of filling of Na$^+$-MMT:P MAG 80:20 and Na$^+$-MMT:PMAG 50:50 (by weight). The offset peak in the small-angle region 2θ=5.0 (d = 1.76 nm) in comparison with the Na$^+$–MMT while filling degree of 20 wt. PMAG% (Fig. 7.2, curve 4) shows that the intercalated nanostructure formed. No peak at higher degrees of filling PMAG 50 wt% (Fig. 7.2, curve 3) indicates that the bundle and eksfolirovanie elementary packages that usually observed at d > 8 to 10 nm [1].

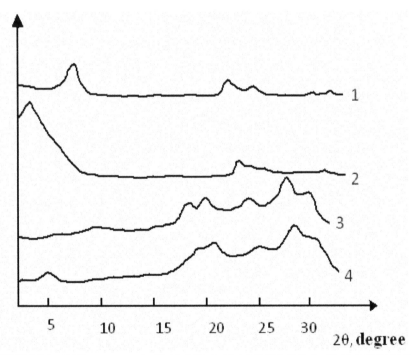

FIGURE 7.2 Diffractograms: 1 – Na$^+$-MMT; 2 – Na$^+$-MMT:MAG; 3 – Na$^+$-MMT:PMAG (80:20); 4 – Na$^+$-MMT:PMAG (50:50)

From the comparison of data on weight loss and DTA composites re-sidual water adsorbed on the external and internal (in interlayer galleries) basal surfaces of Na$^+$–MMT was determined and number PMAG incorporated in composites was determined: the more mass% MAG in the initial aqueous slurry, the greater filling PMAG interlayer galleries Na⁺–MMT (Table 7.1).

TABLE 7.1   The Ratio of Components in the Composites

The ratio Na⁺-MMT:MAG (% by weight) in the initial suspension	Filling PMAG in wt.% of the total weight of the nanocomposite Na⁺-MMT: PMAG
30:70	35.7
40:60	24.6
50:50	12.6

Study of the sorption capacity Na⁺–MMT nanocomposites: PAG/Na⁺-MMT: PMAG and Ca^{2+} MTT:MTT PAG/Ca^{2+}:PMAG using the model aqueous solutions of metal salts of lead, cadmium, zinc, copper, cobalt, chromium (VI), molybdenum and tungsten have shown that they have a high sorption activity in relation to heavy metal ions. Oil capacity of nano composites was also determined, so for the Ca^{2+}-MMT:PMAG (50:50), it is in static conditions 4.5 g of oil/1 g of sorbent in a dynamic oil 20–50 g/1 g of sorbent (depending on the composition of a sorbent) sorption capacity for heavy metals 280–600 meq/100 g.

The content of functional groups in the nanocomposites guanidine desired products gave a bactericidal activity. Antimicrobial activity of Na⁺+MMT, MMT Ca^{2+} + MMT, and organic modifiers nanocomposites showed that they are to varying degrees, have biocidal activity against microorganisms studied. Moreover, nanocomposites, unlike PAG/PMAG have bactericidal activity of all tested strains, while with the increase in MMT (PMAG/PAG) content PMAG/PAG to 40 wt% achieved complete lysis of bacterial cells.

7.3   CONCLUSION

Thus, the new functional nanomaterials combining high bactericidal activity with an increased ability to bind to heavy metals and organic pollutants were received. Certainly, received results broaden the polymer nanocomposites as compared with traditionally discussing areas of their practical use and emphasize the importance of understanding the chemical aspects of the formation of such functional nanomaterials.

KEYWORDS

- **Acrylate guanidine**
- **Methacrylate guanidine**
- **Montmorillonite**
- **Nanocomposites**

REFERENCES

1. Mikitaev, A. K., Kozlov, G. V., & Zaikov, G. E. (2009). *Polymer Nanocomposites: A variety of structural forms and applications*, Moscow: Science, 278 p.

CHAPTER 8

TISSUE DIFFERENTIATION AND PECULIARITIES OF SUGAR BEET (*BETA VULGARIS* L.) PLANTS PLASTIC METABOLISM IN THE PROCESS OF INDIRECT MORPHOGENESIS IN-VITRO CULTURE

O. L. KLYACHENKO, A. F. LIKHANOV, and S. A. KRYLOVSKA

CONTENTS

ABSTRACT

The chapter presents the results of experiments performed for sugar beet (*Beta vulgaris* L.) plant regenerants obtaining of one sort and five hybrids through the indirect morphogenesis. Stages and peculiarities of regenerants' indirect morphogenesis in vitro culture, as well as specific characters of morphogenic structures in sugar beet calluses were studied. The effectiveness of triterpene saponosides usage as diagnostic markers, which determine potentially high productivity and adaptiveness of sugar beet (*Beta vulgaris* L.) plants, is proved.

8.1 INTRODUCTION

In the sugar beet (*Beta vulgaris* L.) breeding methods of plant culture in vitro, which allows growing and investigating plant organisms and tissues in relatively controlled conditions are widely used [5]. One promising direction in breeding of this culture is plant regenerants obtaining through the indirect morphogenesis [15]. It is known that the procedure of the sugar beet calluses obtaining, which have ability of embryogenesis, is composite and has sort specific peculiarities [16]. Culture needs careful selection of nutrient medium, including hormones, vitamins and amino acids [1]. There are reports on the development of modified growth media [12, 15], optimal conditions for the induction of sugar beets embryogenesis and obtaining of viable plant regenerants [1]. However, issues about regulation of morphogenetic processes in calluses, interaction between non-differentiated and highly specialized cells, spatial organization and development of meristemoids are poorly studied. There are conflicting data on the initiation of morphogenic zones in sugar beet calluses from cells of the upper epidermal layers of explants or from the deeper tissues [1]. Efforts are underway to find inducers of adventitious organs morphogenesis (buds, roots) and embryos that have fundamental differences in the tissue structures formation and conducting systems. The role of biopolymers that are part of cell walls in callus cells, intertissue transport of hormones and nutrients, along with the processes of meristemoids formation and initiation of the primary tissue differentiation are not studied completely. An important element in the sugar beet breeding with useful agrochemical and agri-environmental values is identification of biochemical markers that allow making primary screening of plants in vitro conditions. As such, the purpose of our study was to explore peculiarities of the meristemoids formation in sugar beet callus tissue, to determine the specific transformation of cells in morphogenic zones and identify biochemical markers in regenerated plants, which characterize varieties with biological indicators of the prime importance with the possibility of their usage in the primary diagnostics and plant breeding.

8.2 MATERIAL AND METHODS

In investigation diploid sort Yaltushkivsky single-seeded 64, diploid hybrids Ukrainskiy MS 70, Uladovo-Verhnyatskiy MS 37, Uladovo-Veselopodolyanskiy MS 84, Atamansha and triploid hybrid Alexandriya were used. Morphogenic callus was induced on the modified Murashige-Skoog medium supplemented with 100 mg/L meso-Inositol, 0.5 mg/L thiamine, 1 mg/L 6-benzylaminopurine, 0.5 mg/L naphthaleneacetic acid, 0.1 mg/L indoleacetic acid [11].

Callus was cultured in the darkness at + 25 °C [15]. For cytological and histochemical studies morphogenic sugar beet callus, cultured during three passages (9 weeks), was selected. Plant material was fixed for 24 h according to Chamberlain's protocol [8]. Tissue sections were fixed with iron hematoxylin according to Heidenhain's protocol [8]. Callose depositions were found through fluorescence microscopy by using fluorochrome aniline blue (dilution–1:10000) with microscope Axioscope A–1 Carl Zeiss. Callus tissue was fixed with fluorochrome during 30 min in phosphate buffer (pH–12.0), then washed in buffer twice for 5 min.

Identification of secondary metabolites in plant regenerants tissues was made by TLC. Plates with the size 100×150 mm, the sorbent–Kieselgel 60 F254 (Merck) were used. For the phenol carbonic acids and flavonoids identification were used the next solvent systems: chloroform methano-water (70: 30: 4) and chloroform-gelid acetic acid-methanol-water (60: 32: 12: 8). To enhance the fluorescence of substances in ultraviolet light UV (365 nm), plates with chromatograms were treated with 5% EtOH solution of $AlCl_3$ with following heating (5 min at 105°C). Saponins were detected by the sequential treatment of plates with alcoholic solution of sulfuric acid and vanillin Ref. [10]. Chromatogram was held for 7–10 min at 110°C until the indicative spots appearance. Photographic materials and digital experimental data processing were made in Axio Vision 40 V Carl Zeiss. Digital data processing was made in the program Image Pro Premier 9.0 (Evaluation version). Analysis of chromatograms was performed in the program Sorbflil TLC.

8.3   RESULTS AND DISCUSSION

Investigation of the *Beta vulgaris* L. plant regenerants obtaining process through the indirect morphogenesis in vitro allows to distinguish several successive stages: initiation of callus formation, formation of non-morphogenic callus, initial differentiation of callus cells, formation of primary tissue barriers, which form additional conditions in the intercellular redistribution of nutrients and products of plastic metabolism, formation of morphogenic modules, meristemoids and adventitious buds, development of axillary, formation of plant regenerants with independent root system. Investigation of callus tissues showed that the most compact and structured were cells, which were surrounded with large sized (150–180 microns) nuclear-free parenchyma cells with significant deposits of callose and lignin components in secondary cell walls. Under optimal cultivation conditions, small dense plasma cell initiated formation of meristemoids.

In morphogenic sugar beet callus functionally active meristemoids can be formed on the surface of callus and in the deeper layers of tissue (Fig. 8.1b). Nevertheless formation of a new gradated cell structures not always ends with the development of axial organs, gemma or rhizogenesis. Primary meristematic zone of morphogenic callus usually consist of a small group of 4–8 small (12–15 microns) tightly aligned cells, which development depends on the culture medium composition, the spatial position (contact with the nutrient medium, adaxial or abaxial explants side), functional status and constitutional properties of cells that surround them. Plane and intensity of initial cell division cause the initial morphology of a newly formed structure, which eventually became an active synthesis center of hormones and bioactive substances and thereon began to function as a regulatory center, which synchronizes the division and differentiation of cells. When cells in morphogenic structures differentiate by the vascular type *hydrocites* highly elongated cells are formed, in which gradually formed mesh and spiral thickenings of secondary cell walls [2]. The degree of the cells stretching depended on their spatial position and decreased in the centrifugal direction. The process of the cell walls stretching is known to be regulated by hormones, mainly auxin [9] or its synthetic analogs, that is why according to intensity and direction of cells elongation, specific symplast growth, centers of hormones synthesis and growth factors concentration gradient can be determined (Fig. 8.1a). In peripheral areas of sugar beet callus of Atamansha hybrid parenchyma cells were characterized by a significant increase in size indicators, but unlike provascular chords cells this process was relatively in all directions. Meristematic cones that function as attractants and centers of hormones synthesis, surrounded by lignificated parenchyma envelopes, create conditions for cell differentiation by vascular type (Fig. 8.1b).

Thus, in the callus cells culture, the major structural elements of the axial organ are gradually highlighted. In the most investigated samples of a sugar beet callus tissues small cells with dense cytoplasm were surrounded by a large cells with large deposits of β-(1–3)-glucan (callose) in cell walls (Fig. 8.1c, d).

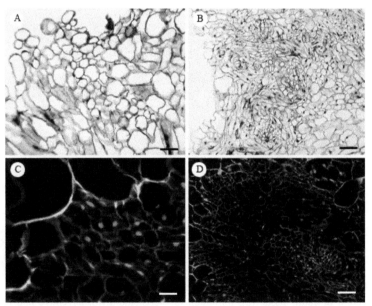

FIGURE 8.1 Differentiation of cells in sugar beet callus tissues in vitro: (a) initial cells differentiation in non-morphogenic callus; (b) cells differentiation with vascular bundle formation; (c) callose deposition (green fluorescence) in peripheral zones of non-morphogenic callus; (d) structuring of callus tissue with highlighted initial morphogenic zones, surrounded with lignified cells (line, C – 25 μm, B, D – 100 μm).

It is known that callose in plant organism performs protective and regulatory functions [4]. Its synthesis is induced increase in the total pool of calcium ions in the cytoplasm of plant cells [14], action of elicitors [4], mechanical effects [13]. Deposition of callose as a dynamic component of the plants cell wall, regulate transport of assimilates in tissues, create conditions for partial or complete isolation of protoplasts from cells from external factors [6]. Polyfunctional polysaccharide creates preconditions in sugar beet callus for selective transport and gradient redistribution of organic substances, products o f primary and secondary metabolism, and as a result causes histochemical heterogeneity of tissues.

In the third callus passage (8–9 week) of Uladovo-Verhniatskiy MS 37 sugar beet cultivar, micro sprouts with fully formed auxiliary meristem typical of this type of structure are formed (Fig. 8.2, b–d).

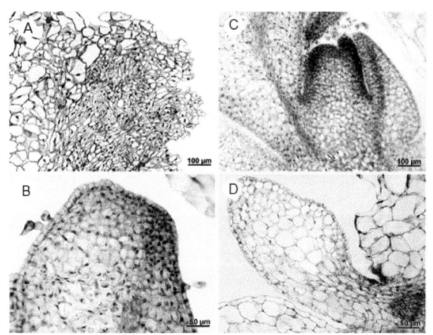

FIGURE 8.2   Formation of sugar beet meristemoids and adventive buds in vitro: module organization of callus tissue (a), formation of meristemoid (b), adventive bud (c) and microsprout (d).

According to the nucleo cytoplasmic index, cytoplasm density, morphology and gradated zonal distribution of cell, apical meristem differs significantly from meristemoids. In meristem cells from tunics to column, size of cells has a distinct trend of the morphological parameters increasing. Cells of apical meristemoids are determined by the less expressed dependence of their size on the morphogenic tissues location in the structure It is possible that in merystemoids cones and small groups of cells that function as attractants and synthesis centers of hormones in the parenchyma surrounded by huge deposits of callose in cell walls starts a cascade of molecular and genetic processes responsible for the regulation of vascular system and gradual differentiation of tissues, which are the main structural elements of the axial organ.

For spatial structural organization morphogenic zone in the callus tissue of sugar beet can be considered as separate modules (Fig. 8.2a). Each of these modules partially or completely separated from the non-morphogenic callus area with large parenchyma cells with developed secondary cell walls. Further realization of structures' morphogenic potential, which formed de novo, depends on cultivation conditions, as well as on the functional activity of meristematic zones. Under optimal conditions in zones of morphogenic modules adventitious buds, from which mature plant regenerants are formed.

In vitro in vegetative parts of sugar beet plant regenerants accumulation of secondary metabolites, including phenolcarbonic acids, flavonoids and saponins, which play an important role in the formation of the constitutional stability, are triggered [3]. General condition of the plant organism and its survival strategy depends on the activity of phenolic compounds and saponins synthesis. TLC in MeOH leaves extract revealed phenol carbonic acid (Fig. 8.3, a and b). By the nature of the fluorescence in the UV (365 nm), the color reaction of $FeCl_3$ solution and R_f the substance the most closely corresponds to chlorogenic acid (Rf = 0.05).

In the chloroform methanol water (70: 30: 4) solvent system after the plate treatment with 5% EtOH solution of $AlCl_3$ we have detected flavonoids and other polyphenolic substances with blue and blue-green fluorescence. It was also marked that in vitro conditions the most active synthesis of aromatic compounds was detected in Uladovo-Verhniatskiy MS 37, Uladovo-Veselopodolyanskiy MS 84 and Alexandriya varieties (sample number 2, 4, 5) and the lowest content of phenolic compounds was found in leaves of Yaltushkivskiy single-seeded 64 (number 1).

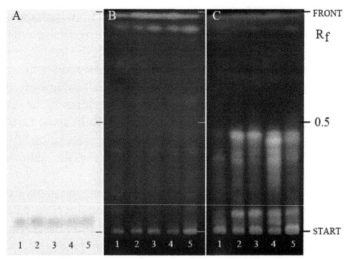

FIGURE 8.3 Chromatogram of sugar beet leaves menthol extracts in vitro (solvent systems: chloroform-methanol-water–70: 30: 4): A – demonstration of simple phenol compounds with 10% $FeCl_3$ solution; B – auto fluorescence of phenol compounds and chlorophylls (a and b); C – fluorescence of phenols after treatment with EtOH solution of $AlCl_3$

Better separation of phenolic compounds in sugar beet tissues showed solvent system: chloroform gelid acetic acid methanol water (60: 32: 12: 8). In these conditions it is possible to divide more than 10 aromatic compounds in the R_f range from 0.1 to 0.72. After chromatograms treatment with 5% EtOH sulfuric acid solution and 1% vanillin solution in the plate purple-violet adsorbent-coated glass strip appeared, which is character for saponins (Fig. 8.4a). Leaves of sugar beet plant regenerants of Ukrainian selection of each selected variety we have detected saponins with R_f indexes 0.21; 0.32; 0.35. In addition in leaves of plant regenerants of Uladovo-Verhniatskiy MS 37 and Atamansha hybrids we have identified compounds with $R_f = 0.56$ and 0.62. In the UV with wavelength 365 nm substance with $R_f = 0.62$ and 0.67 had a bright turquoise color luminescence (Fig. 8.4). Bon the basis of the published data, an organic compound with $R_f = 0.62$ can be defined as oleanolic acid basic derivative substance for the synthesis of sugar beet triterpene saponins [7]. In contrast, a spot with $R_f = 0.56$ strongly absorbed UV (365 nm).

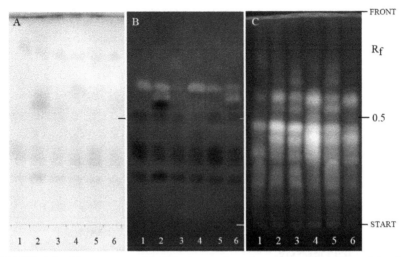

FIGURE 8.4 Chromatogram MeOH of sugar beet leaves extracts cultivated in vitro (chloroform gelid acetic acid methanol water 60: 32: 12: 8): A – saponins identification (revealing agent–5% EtOH solution of H_2SO_4 and 1% vanillin solution); B – plate in UV (365 nm); C – fluorescence of phenolic compounds after treatment with EtOH solution of $AlCl_3$.

It should be noted that the distinctive feature of grades Uladovo-Verhniatskiy MS 37 and Atamansha hybrids is a high yield and roots sugar content, among other things the last variety has high drought tolerance.

Thus, taking into account absolutely identical composition of nutrient media, the same photo and thermal regime of cultivated in vitro plants, triterpenoid saponins with $R_f = 0.56$ and 0.62 can be biochemical markers for Uladovo-Verhniatskiy MS 37 and Atamansha sugar beet varieties (Table 8.1).

The average content of saponins in leaves extracts of sacchariferous plant regenerants Uladovo-Verhniatskiy MS 37 and drought-tolerant Atamansha variety at the average rate was 2–3 times higher than that in Yaltushkivskiy single-seeded 64. Thus, there is a reason to consider reasonable researches of isolated by us compounds usage in the primary diagnostics and plant regenerates breeding on valuable agro-ecological characteristics, including potentially high drought tolerance.

TABLE 8.1   Chromatographic Separation of Triterpene Saponins of Sugar Beet Plant Regenerants

No	Sugar beet variety	R_f						
1	Yaltushkivsky single-seed-ed 64	0.21	0.32	0.35	0.51	-	-	0.67
2	Uladovo-Verhnyatskiy MS 37	0.21	0.32	0.35	0.51	**0.56**	**0.62**	0.67
3	Ukrainskiy MS 70	0.21	0.32	0.35	0.51	-	-	-
4	Uladovo-Veselopodolyans-kiy MS 84	0.21	0.32	0.35	0.51	-	-	0.67
5	Alexandriya	0.21	0.32	0.35	0.51	-	-	0.67
6	Atamansha	0.21	0.32	0.35	0.51	**0.56**	**0.62**	0.67

8.4   CONCLUSIONS

1. Indirect morphogenesis of sugar beet plant in vitro culture pass through several successive stages: initiation of callus formation, formation of non-morphogenic callus, initial differentiation of callus cells, formation of primary tissue barriers, which form additional conditions in the intercellular redistribution of nutrients and products of plastic metabolism, formation of morphogenic modules, meristemoids and adventitious buds, development of axillary, formation of plant regenerants with independent root system.

2. A critical stage morphogenic structures in sugar beet calluses in vitro culture id formation of specific morphogenic modules represented with meristemoids cone (or group of cells with high proliferative activity), provascular zone with hydrocites system and several layers of small parenchyma cells. Each morphogenic module is surrounded by large cells with large deposits of callose, lignin and suberin on cell walls.

3. Was determined that at the stage of sugar beet plant regenerants formation in vitro culture, triterpene saponosides with $R_f = 0.56$ and 0.62 can be an important diagnostic markers in the solvent system: chloroform gelid acetic acid methanol water (60: 32: 12: 8).

KEYWORDS

- **Callus**
- **Explants**
- **Markers**
- **Morphogenesis**
- **Plant regenerants**
- **Sugar beet**
- **Triterpene saponosides**

REFERENCES

1. Bannikova, M. A., Golovko, A. E., Hvedinich, O. A., & Kuchuk, N. V. (1995). *Cytology and Genetics J*, 6, 14 (in Russian).
2. Barikina, R. P., & Churikova, O. A. (2004). *Moscow Univ. Reporter, Biology ser*, 2, 23 (in Russian).
3. Boeva, S. A., Brezhneva, T. A., & Malceva, A. A., et al. (2007). *VNU Reporter, Chemistry. Biology. Pharmacy ser*, 1, 139 (in Russian).
4. Gorshkova, T. A. (2007). *Plant cell wall as a dynamic system*. Moscow 429 p. (in Russian).
5. Kruglova, N. N., Batigina, T. B., & Gorbunova, T. J. (2005). *Embryological basics of wheat androclinium: atlas*. Moscow. 99 p. (in Russian).
6. Kursanov, A. L. (1976). *Transport of assimilates in plants*. Moscow, 825 p. (in Russian).
7. Mironenko, N. V., Brezhneva, T. A., & Selemenev, V. F. (2011). *Chemistry of a plant raw material*, 3, 153 (in Russian).
8. Patusheva, Z. P. (1988). *Plant cytology laboratory manual*. Moscow 271 p. (in Russian).
9. Polevoy, V. V. (1989). *Plant physiology*. Moscow. 464 p. (in Russian).
10. Ladugina, E. Ya., Safronich, L. N., & Otryashenkova, V. E., et al, (1983). *Chemical analysis of herbs*. Moscow, 176 p. (in Russian).
11. Golovko, A., (2001). *Cytology and Genetics J.*, 6, 10 (in Russian).
12. Gurel, S., Gurel, E., & Kaya, Z. (2001). *Turkish Journal of Botany*, 25(1), 25.
13. Jaffe, M. J., Huberman, M., Johnson, J., & Telewski, F. W. (1985). *Physiol. Plant*, 64, 271.
14. Kauss, H. J., (1985). *Cell Sci. Suppl.*, 2, 89.
15. Mishutkina, Ya. V., & Gaponenko A. K. (2006). *Russian Journal of Genetics*, 42(2), 150.
16. Murashige, T., & Skoog, F., (1962). *Physiol. Plant*, 15, 473.

CHAPTER 9

TRANSPORT OF MEDICINAL SUBSTANCES BY CHITOSAN FILMS

A. S. SHURSHINA, E. I. KULISH, V. V. CHERNOVA, and
V. P. ZAHAROV

CONTENTS

ABSTRACT

Sorption and diffusive properties of films are studied. Kinetic curves of release of medicinal substances, having abnormal character are shown. The analysis of the obtained data showed that a reason for rejection of regularities of process of transport of medicinal substance from chitosan films from the classical fikovsky mechanism are structural changes in a polymer matrix, including owing to its chemical modification at interaction with medicinal substance.

9.1 INTRODUCTION

In the last decades intensive researches on development of polymeric systems with controlled delivery of pharmaceuticals as use for this purpose of polymers eliminates many defects of traditional medicinal forms are conducted [1, 2]. Research of regularities of processes of diffusion of water and medicinal substance in polymer films and opportunities of control of release of medicines became the purpose of this work. As a matrix for the immobilization of drugs used naturally occurring polysaccharide chitosan, which has a number of valuable properties: nontoxicity, biocompatibility, high physiological activity [3], as well as a drug used gentamicin and cefatoksim, actively applied in the treatment of pyogenic infections of the skin and soft tissue [4].

9.2 EXPERIMENTAL PART

The object of investigation was a chitosan (ChT) specimen produced by the company "Bioprogress" (Russia) and obtained by acetic deacetylation of crab chitin (degree of deacetylation ~84%) with M_{sd}=334,000. As the medicinal substance (MS) used an antibiotic gentamicin (GM) and cefatoksim (CFT).

To study the release kinetics of MS the sample was placed in a cell with distilled water. Stand out in the aqueous phase MS recorded spectro photo metrically at a wavelength, corresponding to the maximum absorption in the UV spectrum of MS. Quantity of MS released from the film at time t (G_s) was estimated from the calibration curve. The establishment of a constant concentration in the solution of MS G_∞ is the time to equilibrium.

MS mass fraction α, available for diffusion, assessed as the quantity of films released from the antibiotic to its total amount entered in the film. The experimental methods are described in Ref. [5].

9.3   RESULTS AND DISCUSSION

It is well known that release of MS from polymer systems proceeds as diffusive process [6–8]. However a necessary condition of diffusive transport of MS from a polymer matrix is its swelling in water, that is, effective diffusion of water in a polymer matrix. Diffusing in a polymer matrix, water molecules, possessing it is considerable bigger mobility in comparison with high-molecular substance, penetrate in a polymer material, separating apart chains and increasing the free volume of a sample. The main mechanisms in water transport in polymer films are simple diffusion and the relaxation phenomena in swelling polymer. If transfer is caused mainly mentioned processes, the kinetics of swelling of a film is described by the equation [9].

$$m_t/m_\infty = kt^n,\qquad(1)$$

where m_∞ – relative amount of water in equilibrium swelling film sample, k – a constant connected with parameters of interaction polymer-diffuse substance, n – an indicator characterizing the mechanism of transfer of substance. If transport of substance is carried out on the diffusive mechanism, the indicator of n has to be close to 0.5. If transfer of substance is limited by the relaxation phenomena – $n > 0.5$.

The parameter n determined for a film of pure ChT is equal 0.63 (i.e., > 0.5) that is characteristic for the polymers, being lower than vitrification temperature [10]. This fact is connected with slowness of relaxation processes in glassy polymers. Values of equilibrium sorption of water and indicator n defined for film samples, passed isothermal annealing (a relaxation of nonequilibrium conformations of chains with reduction of free volume), are presented in Table 9.1.

TABLE 9.1   Parameters of Swelling of Chitosan Films in Water Vapor

Composition of the film	The concentration of MS in the film, mol/ mol ChT	Annealing time, min	n	Q_∞, g/g ChT
ChT		15	0.50	2.50
		30	0.48	2.48
		60	0.44	2.47
		120	0.43	2.46
ChT-GM	0.01	30	0.36	1.63
		60	0.29	1.54
		120	0.26	1.40
	0.05	30	0.31	1.41
	0.1	30	0.28	1.15
ChT-CFT	0.01	30	0.43	2.20
		60	0.41	1.33
		120	0.40	1.32
	0.05	30	0.41	1.30
	0.1	30	0.39	1.05

Apparently from Table 9.1 data, carrying out isothermal annealing leads to that values of an indicator n decrease. Thus, if annealing was carried out during small time (15–30 min), the value n determined for pure ChT is close to 0.5. It indicates that transfer of water is limited by diffusion, and it is evidence that ChT in heat films is in conformational relaxed condition. In process of increase time of heating till 60–120 min, values of an indicator n continue to decrease that, most likely, reflects process of further restructuring of the polymer matrix, occurring in the course of film heating. That processes of isothermal annealing of ChT at temperatures $\geq 100°C$ are accompanied by course of a number of chemical transformations, was repeatedly noted in literature [11, 12]. In particular, it is revealed that besides acylation reaction, there is the partial destruction of polymer increasing the maintenance of terminal aldehyde groups which reacting with amino groups, sew ChT macromolecules at the expense of formation of azomethine connections. In Ref. [13], the fact of cross-linking in the ChT during isothermal annealing was confirmed by the study of the spin-lattice relaxation. In the values of equilibrium sorption isothermal annealing, however, actually no clue, probably owing to the low density of cross-links.

A similar result - reduction indicator n is achieved when incorporated into a polymer matrix MS. As the data in Table 9.1, the larger MS entered into the film, the slower and less absorb water ChT. During isothermal annealing medicinal films effect enhanced. Such deviations from the laws of simple diffusion (Fick's law) and others researches have observed, explaining their strong interaction polymer with MS [14].

In the aqueous environment from ChT film with antibiotic towards to the water flow moving to volume of the chitosan, from a polymeric film MS stream is directed to water.

In Fig. 9.1, typical experimental curves of an exit of CFT from chitosan films with different contents of MS are presented. All the kinetic curves are located on obviously expressed limit corresponding to an equilibrium exit of MS (G_∞).

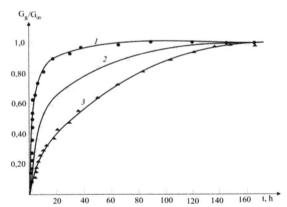

FIGURE 9.1   Kinetic curves of the release of the MS from film systems ChT-CFT with the molar ratio of 1:0.01 (1), 1:0.05 (2) and 1:0.1 (3). Isothermal annealing time – 30 min.

At diffusion MS from films, also as well as in case of sorption by films ChT–MS of water vapor, have anomalously low values of the parameter n. Table 9.2 estimated in this case from the slope in the coordinates $\lg(G_s/G_\infty)$–$\lg t$. Increase the concentration of MS and time of isothermal annealing, as well as in the process of sorption of water vapor films, accompanied by an additional decrease in the parameter n. Symbiotically index n also changes magnitude α. Moreover, the relationship between the parameters of water sorption films ChT-MS (values of diffusion coefficient, an indicator n and values of equilibrium sorption Q) and the corresponding parameters of the diffusion of MS from the polymer film on condition $G_s/G_\infty \leq 0.5$ is shown.

TABLE 9.2 Desorption Parameters GM and CFT in Films Based On ChT

Composition of the film	The concentration of MS in the film, mol/mol ChT	Annealing time, min	n	α
ChT-GM	1:0.01	30	0.31	0.97
		60	0.24	0.92
		120	0.21	0.88
	1:0.05	30	0.20	0.84
	1:0.1	30	0.30	0.88
ChT-CFT	1:0.01	30	0.38	0.97
		60	0.36	0.95
		120	0.31	0.94
	1:0.05	30	0.32	0.94
	1:0.1	30	0.28	0.92

All known types of anomalies of diffusion, can be described within relaxation model [15]. Unlike the fikovsky diffusion assuming instant establishment and changes in the surface concentration of sorbate, the relaxation model assumes change in the concentration in the surface layer on the first-order equation [16]. One of the main reasons causing change of boundary conditions are called nonequilibrium of the structural-morphological organization of a polymer matrix [15]. A possible cause of this could be the interaction between the polymer and MS.

Abnormally low values n in work [17] were explained with presence tightly linked structure of amorphous-crystalline matrix. In this case, is believed, the effective diffusion coefficient in process of penetration into the volume of a sample can be reduced due to steric constraints that force diffusant bypass crystalline regions and diffuse to the amorphous mass of high-density cross-linking. As ChT belongs to the amorphous-crystalline polymers, it would be possible to explain low values n observed in our case similarly. However, according to Table 9.2 in the case of films of individual ChT such anomalies are not observed. Thus, the effect of substantially reducing n is associated with the interaction between ChT and MS.

IR-and UV-spectroscopy data indicate to taking place interaction between MS and ChT. Binding energy in the adduct reaction ChT–antibiotic, estimated by the shift in the UV spectra of the order of 10 kJ/mol, which allows to tell about connection ChT–antibiotic by hydrogen bonds.

Thus, MS may interact with ChT by forming hydrogen bonds. However, interpretation of the data on the diffusion is much more important that GM can form chains linking ChT by salt formation. Due dibasic sulfuric acid, it is possible to suggest the formation of two types of salts, providing stapling ChT macromolecules with the loss of its solubility. Firstly, the water-insoluble "double" salt - sulfate ChT-GMS, secondly, the salt mixture - insoluble in water ChT sulfate and soluble GM acetate. If to take the CFT, an exchange reaction between ChT acetate and CFT reduces the formation of dissociated soluble salts. Accordingly, the reaction product in this case will consist of the H-complex ChT-CFT.

Data on a share of antibiotic related to polymer adducts (β), obtained in solutions of acetic acid, are presented in Table 9.3.

TABLE 9.3 Mass Fraction of the Antibiotic β, defined in Reaction Adducts Obtained from 1% Acetic Acid

Used antibiotic	The concentration of MS in the film, mol/mol ChT	β
GM	1.00	0.69
	0.10	0.30
	0.05	0.20
	0.01	0.06
CFT	1.00	0.16
	0.10	0.09
	0.05	0.05
	0.01	0.02

As seen in Table 9.3, from the fact that GM is able to "sew" chitosan chain is significantly more closely associated with macromolecules MS than for CFT. Formation of chemical compounds of MS with ChT is probably the reason for the observed anomalies – reducing the rate of release of MS from film caused by simple diffusion, as well as the reduction of the share allocated to the drug (α). Indeed, the proportion of MS found in adducts of reaction correlates with the share of the antibiotic is not capable of participating in the diffusion process, and with the index n, reflecting the diffusion mechanism.

Thus, structural changes in the polymer matrix, including as a result of its chemical modification of the interaction with the drug substance, cause deviations regularities of transport MS of chitosan films from classic fikovskogo mechanism. Mild chemical modification, for example by cross-linking macromolecules salt formation, not affecting the chemical structure of the drug, is a possible area of control of the transport properties of medicinal chitosan films.

KEYWORDS

- **Chitosan**
- **Diffusion**
- **Medicinal substance**
- **Sorption**

REFERENCES

1. Shtilman, M. I. (2006). Polimeryi medico-biologicheskogo naznacheniya, M. Akademkniga, 58–59.
2. Plate, N. A., & Vasilev, A. E. (1986). Fiziologicheski aktivnyie polimeryi. M. Himiya, 152.
3. Skryabin, K. G., Vihoreva, G. A., & Varlamov, V. P. (2002). Hitin I hitozan. Poluchenie, svoistva i primenenie. Nauka, M., 365.
4. Mashkovskii, M. D. (1997). Lekarstvennyie sredstva. Harkov: Torsing 278.
5. Kulish, E. I., Shurshina, A. S., & Kolesov, S. V. (2013). Russian Journal of Applied Chemistry 10 1537–1544.
6. Ainaoui, A., & Verganaud, J. M. (2000). Comput Theor. polym. Sci. 2, 383.
7. Kwon, J. H., Wuethrich, T., Mayer, P., & Escher B.I. (2009). Chemosphere. 76, 83.

8. Martinelli, A., D¢Ilario, L., Francolini, I., & Piozzi, A. (2011). Int. J. Pharm. 1–2, 197.
9. Hall, P.J., & Thomas, K. M. (1992). Fuel, 11, 1271.
10. Chalyih Diffuziya, A. E. (1987). v polimernyih sistemah. M. Himiya, 136.
11. Zotkin, M. A., Vihoreva, G. A., & Ageev, E. P (2004). Himicheskaya tehnologiya 9(15).
12. Ageev, E. P., Vihoreva, G. A., & Zotkin, M. A. (2004). Vyisokomolekulyarnyie soedineniya 12., 2035.
13. Smotrina, T. V. (2012). Butlerovskie soobscheniya. 12, 98–101.
14. Singh, B., & Chauhan, N. (2008). Acta Bio materialia. 1, 1244.
15. Ya. A., & Malkin, A. E. (1979). Chalyih Diffuziya i vyazkost polimerov. Metodyi izmereniya. M. Himiya, 304.
16. Pomerancev, A. L. (2003). "Metodyi nelineinogo regressionnogo analiza dlya modelirovaniya kinetiki himicheskih I fizicheskih processov." Ph.D. Thesis, Moscow.
17. Kuznecov, P. N., Kuznecova, L. I., & Kolesnikova, S. M. (2010). Himiya v interesah ustoichivogo razvitiya, 18, 283–298.

CHAPTER 10

MICROHETEROGENEOUS ND-BASED ZIEGLER-NATTA CATALYST: NEW WAY OF INCREASE ACTIVITY

VADIM Z. MINGALEEV, ELENA M. ZAKHAROVA, and
VADIM P. ZAKHAROV

CONTENTS

ABSTRACT

New way to increase activity of neodymium catalyst in the isoprene polymerization is shown. There exists possibility of increasing activity by
hydrodynamic effect in turbulent flows at the stage of synthesis isopropanol complex with neodymium chloride. In this case the increase in the
content of isopropanol in the complex and decrease the size of its particles.
As a result the catalyst complex forms a stable in time and high activity.
When the polymerization of isoprene on the catalyst polymer has such a
narrow MWD.
PACS: 82.65+*r*, 82.35–*x*.

10.1 INTRODUCTION

Catalyst systems based on lanthanide compounds, particularly neodymium, are among the most studied objects stereospecific polymerization under the action of the Ziegler-Natta catalyst. Currently synthesized many
new lanthanide complexes that are active in the polymerization of dienes
to form stereoregular polymers of different composition [1]. The most
popular practice are microheterogeneous catalysts $NdCl_3$. For example,
such catalysts are used in the synthesis of polyisoprene rubber, which contains up to 98−99% of cis-1,4-units [2].

  The active site formation route in the case of neodymium catalytic systems differs from that in the case of titanium systems. Specifically, the first
stage in the reaction of a neodymium chloride complex with an organo
aluminum compound involves elimination of the ligand (L) from the coordination sphere of neodymium with formation of a solid phase [2]

$$NdCl3 \cdot nL + nAlR3 \rightarrow NdCl3\downarrow + nAlR3 \cdot L. \qquad (1)$$

  Formation of particles of neodymium chloride is due to its strong tendency to form bridging chloride bonds. The elimination of IPA from the
coordination sphere of neodymium is paralleled by alkylation of neodymium chloride:

$$NdCl3 + AlR3 \rightarrow NdCl2R + AlR2Cl. \qquad (2)$$

The resultant neodymium alkyldichloride, both individually and as part of complexes with both AlR3 and its chlorinated derivatives, is the main component of the active site on the particle surface. Obviously, the activity of the sites of stereo specific distribution of the neodymium component of the catalytic system. We showed previously [3] that more vigorous agitation of the reaction mixture in the step of preparation of a neodymium chloride alcohol complex causes enhancement of the catalyst activity in polymerization and narrowing of the molecular weight distribution (MWD) of the polymer. The existing views of the mechanism operating in the stages preceding the formation of active sites in neodymium catalysts associate these effects with the particle size reduction for neodymium chloride isopropanol complex and with the necessary amount of isopropanol it contains. Physically, the initial stages in the reaction of the components of the catalytic system proceed under conditions of maximum homogeneity, that is, with small particles of neodymium chloride complex characterized by $1.95 \leq n < 3$. Increase in the particle size of neodymium chloride isopropanol complex and decrease in n lead to lower activity of the neodymium catalyst in isoprene polymerization and broader MWD of the polymer synthesized. Thus, it is reasonable to carry out modification of the neodymium component of the catalytic system via subjecting to hydrodynamic action under conditions of intense turbulence of flows.

Here, we analyzed the results of pilot tests of a small size tubular turbulent reactor of the diffuser-confuser type in the step of preparation of neodymium chloride isopropanol complex in cis-1,4-polyisoprene production.

10.2 EXPERIMENTAL PART

During the tests, a small-size tubular turbulent reactor of the diffuser-confuser type (diffuser diameter 50 mm, confuser diameter 25 mm, diffuser-confuser section length 150 mm, 8 sections) was installed in the step of preparation of a neodymium chloride alcohol complex (Sintez–Kauchuk Public Joint-Stock Company, Sterlitamak). The design of the reactor was tailored to meet the physicochemical criteria of the process by optimizing the dimensions of the mixing zone for achieving a maximal dissipation of the specific kinetic energy of turbulence [6]. Modification of $NdCl3 \cdot n$IPA complex was achieved via multiple circulation of the suspension through the tubular apparatus installed in the external loop of the volume mixing apparatus (Fig. 10.1). During circulation of the suspension, samples of neodymium chloride isopropanol complex were taken for preparing the NdCl3·nIPA–Al(i-C4H9)3 catalytic complex in the presence of modifying additions of piperylene. The neodymium component of the catalytic complex was prepared from NdCl3·0.6H2O hydrate (hereinafter, NdCl3) with isopropanol in liquid paraffin at 25°C, which procedure yielded a 9 wt.% suspension [4]. The reactant molar ratio was NdCl3 IPA=1:3. The catalytic complex NdCl3·nIPA:Al(i-C4H9):piperylene = 1:13:2.6 was prepared in toluene. Polymerization was carried out in isopentane at isoprene concentration of 16.6 wt%. Polymerization was terminated, and the polymerizate washed, with deionizer water at flow rates of 0.025 and 0.35 m^3 ton^{-1}, respectively. The isoprene conversion achieved under the experimental conditions was 69.4% at the catalyst dosage (on NdCl3 basis) of 1 mol per 24.97×10^3 mol of isoprene.

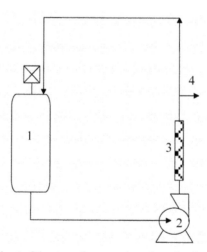

FIGURE 10.1   Flow sheet of the step of preparation of neodymium chloride isopropanol complex. (*1*) Volume apparatus with a stirrer, (*2*) centrifugal pump, (*3*) tubular turbulent apparatus, and (*4*) sampling of suspension for preparation of the catalyst and polymerization.

The catalyst particle size distribution was determined by laser diffraction/scattering on a Sald-7101 (Shimadzu) instrument with the operating measurement range of particle size from 10 nm to 300 μm at the semiconductor laser wavelength of 375 nm.

10.3   RESULTS AND DISCUSSION

Pilot tests showed that multiple circulation of the neodymium chloride alcohol complex suspension caused reduction in the particle size of isopropanol complex (Fig. 10.2). After 40 cycles of circulation through the tubular turbulent apparatus the particle size virtually ceased to decrease. Obviously, during circulation the amount of isopropanol incorporated into the neodymium chloride complex increases, as determined by the $NdCl_3$ particle size.

In the course of circulation of neodymium chloride alcohol complex in the loop without the turbulence apparatus installed the particle size of the suspension decreased from 0.37 to 0.17 μm (Table 10.1), as accompanied by increases in the content of isopropanol in the complex from 1.9 to 2.3 mol $(NdCl3 \ mol)^{-1}$ and in the average molecular weights.

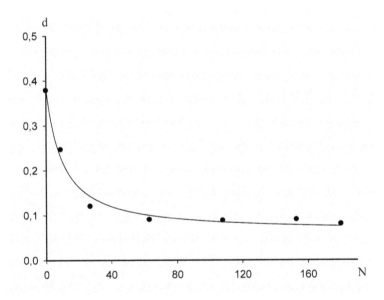

FIGURE 10.2   The $NdCl_3 \cdot nIPA$ particle diameter d, μm, vs. number of cycles of circulation N through the tubular turbulent apparatus. Pumping capacity 6.3 m³/h, average suspension volume 700 L.

TABLE 10.1   Polymerization of Isoprene in the Presence of NdCl3·nIPA–Al(i-C4H9)3–piperylene*

τ_c, h	d_c, μm		n		τ_p, min	U, %		$M_w 10^{-4}$		$M_n 10^{-4}$		M_w/M_n	
	1	2	1	2		1	2	1	2	1	2	1	2
					5	18.3	17.4	54.2	66.5	15.9	27,7	3.4	2.4
1	0.37	0.23	1.90	1.94	40	78.6	79.2	52.8	68.7	11.5	26,4	4.6	2.6
					90	88.1	89.7	57.3	65.4	13.6	24,2	4.2	2.7
					5	17.8	19.3	59.2	68.2	18.5	32,5	3.2	2.1
12	0.26	0.11	2.10	2.55	40	80.2	86.3	64.7	65.8	18.0	28.6	3.6	2.3
					90	93.7	91.2	66.2	67.3	19.5	26.9	3.4	2.5
					5	22.6	25.4	70.1	69.4	22.6	31.5	3.1	2.2
20	0.17	0.087	2.30	2.91	40	89.8	91.7	75.4	77.3	22.8	32.2	3.3	2.4
					90	90.2	94.3	72.6	74.2	20.7	35.3	3.5	2.1

*τ_c is the circulation time for NdCl$_3$·nIPA suspension, d_c is NdCl$_3$·nIPA particle diameter, and τ_p is Polymerization Time; Circulation of the Suspension (*1*) without and (*2*) with a Tubular Turbulent apparatus installed in the loop.

The breadth of the MWD of polyisoprene lies in the Mw/Mn =3.1–4.6 range, being virtually independent of the time of circulation of neodymium chloride isopropanol complex in the external loop. With the tubular turbulent apparatus installed in the loop, the NdCl3 particle size decreases, and the content of isopropanol in the complex, increases. In particular, after 20-h circulation of the suspension through the turbulent apparatus the alcohol content reaches 2.9 mol (NdCl3 mol)$^{-1}$. This allows increasing both the number-and weight-average molecular weights and synthesizing polyisoprene with a narrower MWD (Mw/Mn = 2.1–2.6). The polyisoprene synthesized in the presence of the neodymium catalyst is characterized by a narrow molecular weight distribution.

Solution of the inverse problem of formation of MWD [5] shows that, for isoprene polymerization over NdCl3·nIPA–Al(i-C4H9)3–piperylene catalyst synthesized with the use of complexes containing active sites are formed. In isoprene polymerization catalyzed by the system obtained from a neodymium chloride complex with a low isopropanol content, two types of sites are involved (Fig. 10.3). The polymerization sites are produced by the polymer macromolecules with the most probable molecular weights: type A with ln M = 12.8 and type B with ln M = 14.2. It is seen that, in the 0.75–1.7 L range, the isopropanol content does not affect the number of types of active sites. Further increase in the isopropanol content in the neodymium chloride complex results in formation of a single-site neodymium catalyst, which leads to the polymer with ln M = 13.4 (type C). The functioning in the reaction mixture of one type of macromolecular growth sites results in polyisoprene with a narrow MWD (see Table 10.1).

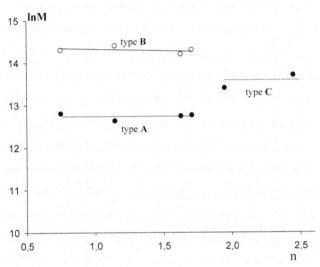

FIGURE 10.3 Average molecular weight M of polyisoprene, corresponding to the individual types of active sites, in relation to the composition of IPA/NdCl$_3$ neodymium chloride isopropanol complex.

10.4 CONCLUSION

Thus, the conditions of formation of neodymium chloride isopropanol complex, in particular, the hydrodynamic mode of the suspension motion, not only significantly affect its composition but also are essential for the functioning of particular isoprene polymerization sites. The reaction of NdCl3·nIPA with Al(i-C4H9)3 at $n\to3$ result in a single-site neodymium catalyst. In terms of the average molecular weights of the polymer synthesized, this type of sites is intermediate between the sites formed in the catalyst based on complexes of NdCl3 with lower content of isopropanol.

Evidently, formation of small particles of neodymium chloride solvate with a high content of alcohol (ligand) result in "averaging" of the structure of the active sites and, consequently, to a smaller difference in the chain termination probability. Since one of the factors responsible for non-uniformity of the active sites in terms of reactivity is the difference in the degree of alkylation of neodymium chloride [2, 5], it can be presumed that the NdCl3 particle size of 100 nm ensures directed alkylation with formation of type C active sites with uniform structure.

The pilot tests showed that the particle size reduction of neodymium chloride alcohol complex in the tubular turbulent apparatus provides high activity of the catalytic system with neodymium component aged for different times. In particular, with the neodymium chloride alcohol complex used for preparation of the catalytic system after aging for 2, 3, and 86 days, the isoprene conversion reaches 85, 80, and 88% in 1 h of polymerization, respectively. For a pilot batch (1027 ton) prepared with a tubular turbulent apparatus installed in the external loop of circulation of neodymium chloride isopropanol complex, the catalyst consumption was 0.98 mol per commercial rubber ton. This level is by over 10% lower than that for the basic procedure of preparation of a neodymium catalytic system.

Thus, we showed that one of the most important preconditions to formation of highly active single-site neodymium catalyst for synthesis of cis-1,4-polyisoprene with a narrow MWD is the use of a suspension of neodymium chloride solvate with the particle size of ca. 100 nm and isopropanol content of up to 3 mol NdCl3 mol^{-1}. This necessitates modification of the neodymium chloride solvate suspension in turbulent flows as a preparatory stage to the reaction with the organo aluminum component. A tubular turbulent reactor of the diffuser-confuser type, in which the neodymium chloride suspension is subjected to multiple hydrodynamic actions, ensures close to homogeneous conditions for formation of the main fragments of the active site.

10.5   ACKNOWLEDGMENTS

This study was financially supported by the Council of the President of the Russian Federation for Young Scientists and Leading Scientific Schools Supporting Grants (project no. MD-4973.2014.8).

KEYWORDS

- Active site
- ND-based catalyst
- Polyisoprene
- Turbulent reactor

REFERENCES

1. Zhang, Z., Cui, D., Wang, B., et al., (2010). Struct. Bond 137, 49–108.
2. Friebe, L., Nuyken, O., & Obrecht, W. (2006). Neodymium Based Ziegler Catalysts Fundamental Chemistry in Adv. Polym. Sci. edited by O. Nuyken, Berlin Heidelberg New York: Springer 296 p.
3. Zakharov, V. P., Mingaleev, V. Z., Morozov, Yu. V., et al. (2012). Russian Journal of Applied Chemistry 85, 965–968.
4. Patent, R., (2011). U 2468995.
5. Monakov, Y. B., Sigaeva, N. N., & Urazbaev, V. N. (2005). Active Sites of Polymerization Multiplicity: Stereo specific and Kinetic Heterogeneity, Leiden: Brill Academic, 397 p.

CHAPTER 11

RESEARCH NOTES ON QUANTUM CHEMICAL CALCULATION

V. A. BABKIN, A. A. DENISOV, I. I. BAHOLDIN, and G. E. ZAIKOV

CONTENTS

11.1 THEORETICAL ESTIMATION OF ACID FORCE OF MOLECULE ALLYLBENZOL BY METHOD AB INITIO.

11.1.1 SUMMARY

For the first time it is executed quantum chemical calculation of a molecule of allylbenzol method AB INITIO with optimization of geometry on all parameters. The optimized geometrical and electronic structure of this compound is received. Acid power of allylbenzol is theoretically appreciated. It is established, than it to relate to a class of very weak H–acids (pKa=+33, where pKa – universal index of acidity).

11.1.2 AIMS AND BACKGROUNDS

The Aim of this work is a study of electronic structure of molecule allylbenzol [1] and theoretical estimation its acid power by quantum-chemical method AB INITIO in base 6–311G**. The calculation was done with optimization of all parameters by standard gradient method built-in in PC GAMESS [2]. The calculation was executed in approach the insulated molecule in gas phase. Program MacMolPlt was used for visual presentation of the model of the molecule [3].

11.1.3 METHODOLOGY

Geometric and electronic structures, general and electronic energies of molecule allylbenzol was received by method AB INITIO in base 6-311G** and are shown on Fig. 11.1 and in Table 11.1. The universal factor of acidity was calculated by formula: pKa = 49.04 – 134.6 q_{max}^{H+} [4, 5], where, q_{max}^{H+} is a maximum positive charge on atom of the hydrogen $q_{max}^{H+} = +0.12$ (for allylbenzol q_{max}^{H+} alike Table 11.1). This same formula is used in references [6–16] (pKa=33).

Quantum-chemical calculation of molecule allylbenzol by method AB INITIO in base 6–311G** was executed for the first time. Optimized geometric and electronic structure of thise compound was received. Acid power of molecule allylbenzol was theoretically evaluated (pKa=33). Thise compound pertain to class of very weak H-acids (pKa>14).

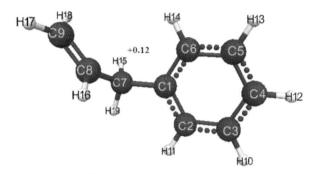

FIGURE 11.1   Geometric and electronic molecule structure of allylbenzol.
($E_0 = -910074$ kDg/mol, $E_{el} = -1945873$ kDg/mol)

TABLE 11.1   Optimized Bond Lengths, Valence Corners and Charges on Atoms of the
Molecule Allylbenzol

Bond lengths	R,A	Valence corners	Grad	Atom	Charges on atoms
C(2)-C(1)	1.39	C(5)-C(6)-C(1)	121	C(1)	−0.13
C(3)-C(2)	1.39	C(1)-C(2)-C(3)	121	C(2)	−0.09
C(4)-C(3)	1.38	C(2)-C(3)-C(4)	120	C(3)	−0.09
C(5)-C(4)	1.39	C(3)-C(4)-C(5)	119	C(4)	−0.10
C(6)-C(5)	1.38	C(4)-C(5)-C(6)	120	C(5)	−0.09
C(6)-C(1)	1.39	C(2)-C(1)-C(6)	118	C(6)	−0.06
C(7)-C(1)	1.52	C(2)-C(1)-C(7)	121	C(7)	−0.11
C(8)-C(7)	1.51	C(1)-C(7)-C(8)	113	C(8)	−0.14
C(9)-C(8)	1.32	C(7)-C(8)-C(9)	125	C(9)	−0.19
H(10)-C(3)	1.08	C(2)-C(3)-H(10)	120	H(10)	+0.09
H(11)-C(2)	1.08	C(1)-C(2)-H(11)	120	H(11)	+0.09
H(12)-C(4)	1.08	C(3)-C(4)-H(12)	120	H(12)	+0.09
H(13)-C(5)	1.08	C(4)-C(5)-H(13)	120	H(13)	+0.10
H(14)-C(6)	1.08	C(5)-C(6)-H(14)	120	H(14)	+0.09
H(15)-C(7)	1.09	C(1)-C(6)-H(14)	119	**H(15)**	**+0.12**
H(16)-C(8)	1.08	C(1)-C(7)-H(15)	110	H(16)	+0.10
H(17)-C(9)	1.08	C(7)-C(8)-H(16)	116	H(17)	+0.11
H(18)-C(9)	1.08	C(8)-C(9)-H(17)	121	H(18)	+0.10
H(19)-C(7)	1.09	C(8)-C(9)-H(18)	122	H(19)	+0.11
		C(1)-C(7)-H(19)	109		

11.2   THEORETICAL ESTIMATION OF ACID FORCE OF MOLECULE 5-METHYLACENAPHTELENE BY METHOD AB INITIO.

11.2.1   SUMMARY

For the first time quantum chemical calculation of a molecule of 5-methylacenaphtelene is executed by method AB INITIO with optimization of geometry on all parameters. The optimized geometrical and electronic structure of this compound is received. Acid power of 5-methylacenaphtelene is theoretically appreciated. It is established, than it to relate to a class of very weak H-acids (pKa = +34, where pKa – universal index of acidity).

11.2.2   AIMS AND BACKGROUNDS

The Aim of this work is a study of electronic structure of molecule 5-methylacenaphtelene [1] and theoretical estimation its acid power by quantum-chemical method AB INITIO in base 6–311G**. The calculation was done with optimization of all parameters by standard gradient method built-in in PC GAMESS [2]. The calculation was executed in approach the insulated molecule in gas phase. Program MacMolPlt was used for visual presentation of the model of the molecule [3].

11.2.3   METHODOLOGY

Geometric and electronic structures, general and electronic energies of molecule 5-methylacenaphtelene was received by method AB INITIO in base 6–311G** and are shown on Fig. 11.2 and in Table.11.2. The universal factor of acidity was calculated by formula: pKa = 49.04–134.6 × q_{max}^{H+} [4, 5], where, q_{max}^{H+} is a maximum positive charge on atom of the hydrogen q_{max}^{H+} = +0.11 (for 5-methylacenaphtelene q_{max}^{H+} alike Table 11.2). This same formula is used in references [6–16] (pKa=34).

Quantum-chemical calculation of molecule 5-methylacenaphtelene by method AB INITIO in base 6–311G** was executed for the first time. Optimized geometric and electronic structure of thise compound was received. Acid power of molecule 5-methylacenaphtelene was theoretically evaluated (pKa=34). Thise compound pertain to class of very weak H-acids (pKa>14).

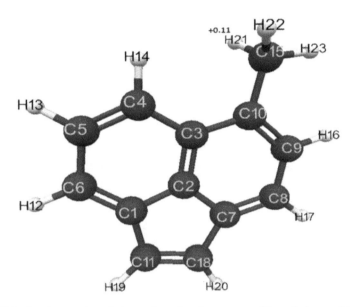

FIGURE 11.2 Geometric and electronic molecule structure of 5-methylacenaphtelene. ($E_0 = -1305744$ kDg/mol, $E_{el} = -3173620$ kDg/mol)

TABLE 11.2 Optimized Bond Lengths, Valence Corners and Charges on Atoms of the Molecule 5-Methylacenaphtelene

Bond lengths	R,A	Valence corners	Grad	Atom	Charges on atoms
C(2)-C(1)	1.41	C(1)-C(2)-C(3)	125	C(1)	−0.04
C(3)-C(2)	1.38	C(9)-C(10)-C(3)	118	C(2)	−0.11
C(3)-C(10)	1.44	C(2)-C(3)-C(4)	116	C(3)	+0.02
C(4)-C(3)	1.42	C(3)-C(4)-C(5)	120	C(4)	−0.08
C(5)-C(4)	1.37	C(1)-C(6)-C(5)	118	C(5)	−0.08
C(5)-C(6)	1.42	C(2)-C(1)-C(6)	119	C(6)	−0.06
C(6)-C(1)	1.36	C(1)-C(2)-C(7)	110	C(7)	−0.04
C(7)-C(2)	1.41	C(11)-C(18)-C(7)	109	C(8)	−0.05
C(7)-C(18)	1.48	C(2)-C(7)-C(8)	118	C(9)	−0.08
C(8)-C(7)	1.36	C(7)-C(8)-C(9)	119	C(10)	−0.11
C(9)-C(8)	1.43	C(8)-C(9)-C(10)	124	C(11)	−0.07
C(10)-C(9)	1.37	C(2)-C(1)-C(11)	106	H(12)	+0.09

TABLE 11.2   *(Continued)*

Bond lengths	R,A	Valence corners	Grad	Atom	Charges on atoms
C(11)-C(1)	1.48	C(1)-C(6)-H(12)	122	H(13)	+0.09
H(12)-C(6)	1.08	C(4)-C(5)-H(13)	119	H(14)	+0.09
H(13)-C(5)	1.08	C(3)-C(4)-H(14)	121	C(15)	−0.18
H(14)-C(4)	1.07	C(9)-C(10)-C(15)	121	H(16)	+0.08
C(15)-C(10)	1.51	C(8)-C(9)-H(16)	118	H(17)	+0.09
H(16)-C(9)	1.08	C(7)-C(8)-H(17)	122	C(18)	−0.07
H(17)-C(8)	1.08	C(1)-C(11)-C(18)	109	H(19)	+0.09
C(18)-C(11)	1.34	C(1)-C(11)-H(19)	125	H(20)	+0.09
H(19)-C(11)	1.07	C(11)-C(18)-H(20)	126	**H(21)**	**+0.11**
H(20)-C(18)	1.07	C(10)-C(15)-H(21)	111	H(22)	+0.11
H(21)-C(15)	1.09	C(10)-C(15)-H(22)	111	H(23)	+0.10
H(22)-C(15)	1.09	C(10)-C(15)-H(23)	111		
H(23)-C(15)	1.08				

11.3   THEORETICAL ESTIMATION OF ACID FORCE OF MOLECULE 9-VINYLANTHRACENE BY METHOD AB INITIO

11.3.1   SUMMARY

For the first time quantum chemical calculation of a molecule of 9-vinyl-anthracene is executed by method AB INITIO with optimization of geometry on all parameters. The optimized geometrical and electronic structure of this compound is received. Acid power of 9-vinylanthracene is theoretically appreciated. It is established, than it to relate to a class of very weak H-acids (pKa = +33, where pKa – universal index of acidity).

11.3.2   AIMS AND BACKGROUNDS

The Aim of this work is a study of electronic structure of molecule 9-vinylanthracene [1] and theoretical estimation its acid power by quantum-chemical method AB INITIO in base 6-311G**. The calculation was done with optimization of all parameters by standard gradient method built-in in

PC GAMESS [2]. The calculation was executed in approach the insulated molecule in gas phase. Program MacMolPlt was used for visual presentation of the model of the molecule [3].

11.3.3 METHODOLOGY

Geometric and electronic structures, general and electronic energies of molecule 9-vinylanthracene was received by method AB INITIO in base 6-311G** and are shown on Fig. 11.3 and in Table 11.3. The universal factor of acidity was calculated by formula: pKa = 49.04 – 134.6 q_{max}^{H+} [4, 5], where, q_{max}^{H+} is a maximum positive charge on atom of the hydrogen q_{max}^{H+} = +0.12 (for 9-vinylanthracene q_{max}^{H+} alike Table 11.3). This same formula is used in references [6–16] (pKa=33).

Quantum-chemical calculation of molecule 9-vinylanthracene by method AB INITIO in base 6-311G** was executed for the first time. Optimized geometric and electronic structure of thise compound was received. Acid power of molecule 19-vinylanthracene was theoretically evaluated (pKa=33). Thise compound pertain to class of very weak H-acids (pKa>14).

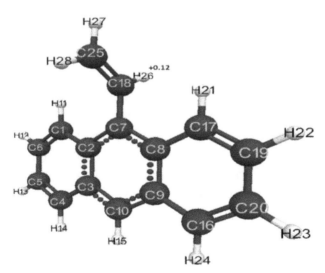

FIGURE 11.3 Geometric and electronic molecule structure of 9-vinylanthracene. (E_0= –1606579 kDg/mol, E_{el}= –4145624 kDg/mol)

TABLE 11.3 Optimized Bond Lengths, Valence Corners and Charges on Atoms of the Molecule 9-Vinylanthracene

Bond lengths	R,A	Valence corners	Grad	Atom	Charges on atoms
C(2)-C(1)	1.44	C(1)-C(2)-C(3)	118	C(1)	−0.05
C(3)-C(2)	1.42	C(2)-C(3)-C(4)	119	C(2)	−0.06
C(4)-C(3)	1.44	C(3)-C(4)-C(5)	121	C(3)	−0.06
C(5)-C(4)	1.34	C(1)-C(6)-C(5)	121	C(4)	−0.06
C(5)-C(6)	1.43	C(2)-C(1)-C(6)	121	C(5)	−0.09
C(6)-C(1)	1.35	C(1)-C(2)-C(7)	123	C(6)	−0.09
C(7)-C(2)	1.40	C(2)-C(7)-C(8)	120	C(7)	+0.08
C(8)-C(7)	1.40	C(7)-C(8)-C(9)	120	C(8)	−0.06
C(9)-C(8)	1.42	C(3)-C(10)-C(9)	121	C(9)	−0.06
C(9)-C(10)	1.39	C(2)-C(3)-C(10)	120	C(10)	−0.01
C(10)-C(3)	1.39	C(2)-C(1)-H(11)	119	H(11)	+0.10
H(11)-C(1)	1.07	C(1)-C(6)-H(12)	120	H(12)	+0.09
H(12)-C(6)	1.08	C(4)-C(5)-H(13)	121	H(13)	+0.09
H(13)-C(5)	1.08	C(3)-C(4)-H(14)	118	H(14)	+0.08
H(14)-C(4)	1.08	C(3)-C(10)-H(15)	119	H(15)	+0.08
H(15)-C(10)	1.08	C(8)-C(9)-C(16)	119	C(16)	−0.06
C(16)-C(9)	1.44	C(7)-C(8)-C(17)	123	C(17)	−0.06
C(17)-C(8)	1.44	C(2)-C(7)-C(18)	120	C(18)	−0.25
C(18)-C(7)	1.50	C(8)-C(17)-C(19)	121	C(19)	−0.08
C(19)-C(17)	1.35	C(16)-C(20)-C(19)	120	C(20)	−0.09
C(19)-C(20)	1.43	C(9)-C(16)-C(20)	121	H(21)	+0.10
C(20)-C(16)	1.35	C(8)-C(17)-H(21)	119	H(22)	+0.09
H(21)-C(17)	1.07	C(17)-C(19)-H(22)	120	H(23)	+0.09
H(22)-C(19)	1.08	C(16)-C(20)-H(23)	121	H(24)	+0.08
H(23)-C(20)	1.08	C(9)-C(16)-H(24)	118	C(25)	−0.14
H(24)-C(16)	1.08	C(7)-C(18)-C(25)	125	**H(26)**	**+0.12**
C(25)-C(18)	1.32	C(7)-C(18)-H(26)	116	H(27)	+0.11
H(26)-C(18)	1.08	C(18)-C(25)-H(27)	121	H(28)	+0.11
H(27)-C(25)	1.08	C(18)-C(25)-H(28)	122		
H(28)-C(25)	1.08				

11.4   THEORETICAL ESTIMATION OF ACID FORCE OF MOLECULE 1-VINYLPYRENE BY METHOD AB INITIO

11.4.1   SUMMARY

For the first time quantum chemical calculation of a molecule of 1-vinyl-pyrene is executed by method AB INITIO with optimization of geometry on all parameters. The optimized geometrical and electronic structure of this compound is received. Acid power of 1-vinylpyrene is theoretically appreciated. It is established, than it to relate to a class of very weak H-acids (pKa=+34, where pKa – universal index of acidity).

11.4.2   AIMS AND BACKGROUNDS

The Aim of this work is a study of electronic structure of molecule 1-vi-nylpyrene [1] and theoretical estimation its acid power by quantum-chem-ical method AB INITIO in base 6–311G**. The calculation was done with optimization of all parameters by standard gradient method built-in in PC GAMESS [2]. The calculation was executed in approach the insulated molecule in gas phase. Program MacMolPlt was used for visual presenta-tion of the model of the molecule [3].

11.4.3   METHODOLOGY

Geometric and electronic structures, general and electronic energies of molecule 1-vinylpyrene was received by method AB INITIO in base 6-311G** and are shown on Fig. 11.4 and in Table 11.4. The universal factor of acidity was calculated by formula: $pKa = 49.04 - 134.6 \ q_{max}^{H+}$ [4, 5], where, q_{max}^{H+} is a maximum positive charge on atom of the hydro-gen $q_{max}^{H+} = +0.11$ (for 1-vinylpyrene q_{max}^{H+} alike Table 11.4). This same formula is used in references [6–16] (pKa=34).

Quantum-chemical calculation of molecule 1-vinylpyrene by method AB INITIO in base 6-311G** was executed for the first time. Optimized geometric and electronic structure of this compound was received. Acid power of molecule 1-vinylpyrene was theoretically evaluated (pKa=34). Thise compound pertain to class of very weak H-acids (pKa>14).

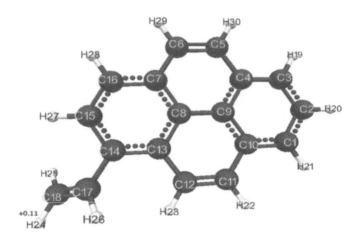

FIGURE 11.4 Geometric and electronic molecule structure of 1-vinylpyrene.
$(E_0 = -1805204 \text{ kDg/mol}, E_{el} = -4860555 \text{ kDg/mol})$

TABLE 11.4 Optimized Bond Lengths, Valence Corners and Charges on Atoms of the Molecule 1-vinylpyrene

Bond lengths	R,A	Valence corners	Grad	Atom	Charges on atoms
C(2)-C(1)	1.38	C(1)-C(2)-C(3)	120	C(1)	−0.06
C(3)-C(2)	1.38	C(2)-C(3)-C(4)	121	C(2)	−0.07
C(4)-C(3)	1.39	C(3)-C(4)-C(5)	122	C(3)	−0.07
C(5)-C(4)	1.45	C(4)-C(5)-C(6)	121	C(4)	−0.03
C(6)-C(5)	1.34	C(5)-C(6)-C(7)	121	C(5)	−0.05
C(7)-C(6)	1.44	C(15)-C(16)-C(7)	121	C(6)	−0.05
C(7)-C(16)	1.39	C(6)-C(7)-C(8)	119	C(7)	−0.02
C(8)-C(7)	1.41	C(4)-C(9)-C(8)	120	C(8)	−0.05
C(8)-C(9)	1.43	C(3)-C(4)-C(9)	119	C(9)	−0.06
C(9)-C(4)	1.41	C(1)-C(10)-C(9)	119	C(10)	−0.03
C(9)-C(10)	1.41	C(2)-C(1)-C(10)	121	C(11)	−0.04
C(10)-C(1)	1.39	C(1)-C(10)-C(11)	122	C(12)	−0.05
C(11)-C(10)	1.44	C(10)-C(11)-C(12)	122	C(13)	−0.03
C(12)-C(11)	1.34	C(8)-C(13)-C(12)	118	C(14)	0.00
C(12)-C(13)	1.45	C(7)-C(8)-C(13)	120	C(15)	−0.06

TABLE 11.4 *(Continued)*

Bond lengths	R,A	Valence corners	Grad	Atom	Charges on atoms
C(13)-C(8)	1.41	C(8)-C(13)-C(14)	120	C(16)	−0.07
C(14)-C(13)	1.40	C(13)-C(14)-C(15)	119	C(17)	−0.17
C(15)-C(14)	1.39	C(14)-C(15)-C(16)	122	C(18)	−0.18
C(16)-C(15)	1.38	C(13)-C(14)-C(17)	121	H(19)	+0.08
C(17)-C(14)	1.49	C(14)-C(17)-C(18)	125	H(20)	+0.09
C(18)-C(17)	1.32	C(2)-C(3)-H(19)	120	H(21)	+0.08
H(19)-C(3)	1.08	C(1)-C(2)-H(20)	120	H(22)	+0.08
H(20)-C(2)	1.08	C(2)-C(1)-H(21)	120	H(23)	+0.09
H(21)-C(1)	1.08	C(10)-C(11)-H(22)	118	**H(24)**	**+0.11**
H(22)-C(11)	1.08	C(11)-C(12)-H(23)	119	H(25)	+0.11
H(23)-C(12)	1.07	C(17)-C(18)-H(24)	121	H(26)	+0.11
H(24)-C(18)	1.08	C(17)-C(18)-H(25)	122	H(27)	+0.09
H(25)-C(18)	1.08	C(14)-C(17)-H(26)	117	H(28)	+0.08
H(26)-C(17)	1.08	C(14)-C(15)-H(27)	119	H(29)	+0.08
H(27)-C(15)	1.07	C(15)-C(16)-H(28)	120	H(30)	+0.08
H(28)-C(16)	1.08	C(5)-C(6)-H(29)	120		
H(29)-C(6)	1.08	C(4)-C(5)-H(30)	118		
H(30)-C(5)	1.08				

11.5 THEORETICAL ESTIMATION OF ACID FORCE OF MOLECULE O-DIVINYLBENZOL BY METHOD AB INITIO

11.5.1 SUMMARY

For the first time quantum chemical calculation of a molecule of o-divinylbenzol is executed by method AB INITIO with optimization of geometry on all parameters. The optimized geometrical and electronic structure of this compound is received. Acid power of o-divinylbenzol is theoretically appreciated. It is established, than it to relate to a class of very weak H-acids (pKa=+34, where pKa – universal index of acidity).

11.5.2 AIMS AND BACKGROUNDS

The aim of this work is a study of electronic structure of molecule o-divinylbenzol [1] and theoretical estimation its acid power by quantum-chemical method AB INITIO in base 6-311G**. The calculation was done with optimization of all parameters by standard gradient method built-in in PC GAMESS [2]. The calculation was executed in approach the insulated molecule in gas phase. Program MacMolPlt was used for visual presentation of the model of the molecule. [3].

11.5.3 METHODOLOGY

Geometric and electronic structures, general and electronic energies of molecule o-divinylbenzol was received by method AB INITIO in base 6-311G** and are shown on Fig. 11.5 and in Table 11.5. The universal factor of acidity was calculated by formula: $pKa = 49.04 - 134.6 \ q_{max}^{H+}$ [4, 5], where, q_{max}^{H+} is a maximum positive charge on atom of the hydrogen $q_{max}^{H+} = +0.11$ (for o-divinylbenzol q_{max}^{H+} alike Table 11.5). This same formula is used in references [6–16] (pKa=34).

Quantum-chemical calculation of molecule o-divinylbenzol by method AB INITIO in base 6-311G** was executed for the first time. Optimized geometric and electronic structure of thise compound was received. Acid power of molecule o-divinylbenzol was theoretically evaluated (pKa=34). Thise compound pertain to class of very weak H-acids (pKa>14).

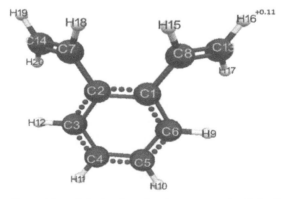

FIGURE 11.5   Geometric and electronic molecule structure of o-divinylbenzol. ($E_0 = -1007846$ kDg/mol, $E_{el} = -2223252$ kDg/mol)

TABLE 11.5 Optimized Bond Lengths, Valence Corners and Charges on Atoms of the Molecule o-divinylbenzol

Bond lengths	R,A	Valence corners	Grad	Atom	Charges on atoms
C(2)-C(1)	1.40	C(1)-C(2)-C(3)	119	C(1)	−0.02
C(3)-C(2)	1.39	C(2)-C(3)-C(4)	121	C(2)	−0.02
C(4)-C(3)	1.38	C(3)-C(4)-C(5)	120	C(3)	−0.08
C(5)-C(4)	1.38	C(4)-C(5)-C(6)	120	C(4)	−0.09
C(6)-C(5)	1.38	C(1)-C(2)-C(7)	121	C(5)	−0.09
C(7)-C(2)	1.49	C(2)-C(1)-C(8)	121	C(6)	−0.08
C(8)-C(1)	1.49	C(5)-C(6)-H(9)	120	C(7)	−0.14
H(9)-C(6)	1.07	C(4)-C(5)-H(10)	120	C(8)	−0.14
H(10)-C(5)	1.08	C(3)-C(4)-H(11)	120	H(9)	+0.10
H(11)-C(4)	1.08	C(2)-C(3)-H(12)	119	H(10)	+0.09
H(12)-C(3)	1.07	C(1)-C(8)-C(13)	125	H(11)	+0.09
C(13)-C(8)	1.32	C(2)-C(7)-C(14)	125	H(12)	+0.10
C(14)-C(7)	1.32	C(1)-C(8)-H(15)	117	C(13)	−0.18
H(15)-C(8)	1.08	C(8)-C(13)-H(16)	121	C(14)	−0.18
H(16)-C(13)	1.08	C(8)-C(13)-H(17)	122	H(15)	+0.10
H(17)-C(13)	1.08	C(2)-C(7)-H(18)	117	**H(16)**	**+0.11**
H(18)-C(7)	1.08	C(7)-C(14)-H(19)	121	H(17)	+0.11
H(19)-C(14)	1.08	C(7)-C(14)-H(20)	122	H(18)	+0.10
H(20)-C(14)	1.08			H(19)	+0.11
				H(20)	+0.11

KEYWORDS

- 1-vinylpyrene
- 9-vinylanthracene
- Acid power
- Allylbenzol
- Method AB initio
- O-divinylbenzol
- Quantum chemical calculation

REFERENCES

1. Kennedi, J. (1978). Cationic polymerization of olefins Moscow. 431 p.
2. Shmidt, M. W., Baldrosge, K. K., Elbert, J. A., Gordon, M. S., Enseh, J. H., Koseki, S., Matsvnaga, N., Nguyen, K. A., SU, S. J., et al. (1993). J. Comput. Chem. 14, 1347-1363.
3. Bode, B. M., &. Gordon, M. S. (1998). J. Mol. Graphics Mod., 16, 133–138.
4. Babkin, V. A., Fedunov, R. G., Minsker, K. S. et al. (2002) Oxidation communication, 25(1), 21–47.
5. Babkin, V. A. et al. (1998). Oxidation communication, 21(4), 454–460.
6. Babkin, V. A., Yu, V., Dmitriev, G., & Zaikov, E. (2012). Geometrical and electronic structure of molecule benzil penicillin by method AB INITIO. In: Quantum-chemical calculations of molecular system as the basis of nanotechnologies in applied quantum chemistry. Volume I. New York, Nova publisher, 7–10.
7. Babkin, V. A., & Tsykanov, A. B. (2012). Geometrical and electronic structure of molecule cellulose by method AB INITIO. In: Quantum-chemical calculations of molecular system as the basis of nanotechnologies in applied quantum chemistry. Volume I. New York, Nova publisher, 31–34.
8. Babkin, V. A., Yu, V., Dmitriev, G., & Zaikov, E. (2012). Geometrical and electronic structure of molecule aniline by method AB INITIO. In: Quantum-chemical calculations of molecular system as the basis of nanotechnologies in applied quantum chemistry. Volume I. New York, Nova publisher, 89–91.
9. Babkin, V. A., Yu, V., Dmitriev, G., & Zaikov, E. (2012). Geometrical and electronic structure of molecule butene-1 by method AB INITIO. In: Quantum-chemical calculations of molecular system as the basis of nanotechnologies in applied quantum chemistry. Volume I. New York, Nova publisher, 109–111.
10. Babkin, V. A., Yu, V., Dmitriev, G., & Zaikov, E. (2012). Geometrical and electronic structure of molecule butene-2 by method AB INITIO. In: Quantum-chemical calculations of molecular system as the basis of nanotechnologies in applied quantum chemistry. Volume I. New York, Nova publisher, 113–115.
11. Babkin, V. A., & Galenkin, V. V. (2012). Geometrical and electronic structure of molecule 3 3-dimethylbutene-1 by method AB INITIO In: Quantum-chemical calculations of molecular system as the basis of nanotechnologies in applied quantum chemistry. Volume I. New York, Nova publisher, 129–131.
12. Babkin, V. A., & Andreev, D. S. (2012) Geometrical and electronic structure of molecule 4, 4-dimethylpentene-1 by method AB INITIO. In: Quantum-chemical calculations of molecular system as the basis of nanotechnologies in applied quantum chemistry. Volume I. New York, Nova publisher, 141–143.
13. Babkin, V. A., & Andreev, D. S. (2012). Geometrical and electronic structure of molecule 4-methylhexene-1 by method AB INITIO. In: Quantum-chemical calculations of molecular system as the basis of nanotechnologies in applied quantum chemistry. Volume I. New York, Nova publisher, 145–147.
14. Babkin, V. A., & Andreev, D. S. (2012). Geometrical and electronic structure of molecule 4-methylpentene-1 by method AB INITIO. In: Quantum-chemical calculations of molecular system as the basis of nanotechnologies in applied quantum chemistry. Volume I. New York, Nova publisher, 149–151.

15. Babkin, V. A., & Andreev, D. S. (2012). Geometrical and electronic structure of molecule isobutylene by method AB INITIO. In: Quantum-chemical calculations of molecular system as the basis of nanotechnologies in applied quantum chemistry. Volume I. New York, Nova publisher, 155–157.
16. Babkin, V. A., & Andreev, D. S. (2012). Geometrical and electronic structure of molecule 2-methylbutene-1 by method AB INITIO. In: Quantum-chemical calculations of molecular system as the basis of nanotechnologies in applied quantum chemistry. Volume I. New York, Nova publisher, 159–161.

CHAPTER 12

TECHNOLOGY OF MANUFACTURING AND APPLICATION OF A NEW PROBIOTIC PREPARATION FOR AGE PRODUCTION

N. V. SVERCHKOVA, N. S. ZASLAVSKAYA,
T. V. ROMANOVSKAYA, E. I. KOLOMIETS,
A. N. MICHALUK, and M. A. KAVRUS

CONTENTS

ABSTRACT

Based on two bacterial strains *Bacillus subtilis* distinguished by high antagonistic activity towards animal pathogens – bacteria of genera *Escherichia, Staphylococcus, Streptococcus* and enzyme activity (protease, cellulase, xylanase), pilot-plant technology of production and application of complex probiotic preparation for increasing biological digestibility of fodder, immune correction and activation of metabolic processes in reared swine has been developed. Test batch of probiotic (500 doses) was fabricated and investigations evaluating its efficiency at stock farms were conducted.

12.1 INTRODUCTION

Stock breeding efficiency is determined by feed resources and especially by quality of fodder. Essential task of fodder production is formulation and introduction into animal rations of feed mixes promoting maximum assimilation of nutrients in the body to secure its vital functions and possessing prophylactic properties.

Increased productivity of pigs and poultry resulting from enhanced digestion of feed may be achieved using new biologically active agents. Nowadays intensive studies are carried out to elaborate and assess application efficiency of fodder probiotic ingredients based on sporulating bacteria of genus *Bacillus*.

Feeding bacteria-antagonists of pathogenic microbiota as constituents of probiotic compositions to farm stock and poultry promotes recovery of intestinal balance, normalizes biological status, immune response, raises vaccination efficiency and ultimately contributes to high profitability of stock farming [1].

One of key approaches in designing probiotic products is search for a rational preparative form because it largely defines efficient use of novel bio-preparations. The optimal in veterinary-sanitary practice is dry form of probiotic ensuring precision of dosage, compactness, handy package, long shelf life [2–4].

Process of manufacturing dry probiotics-includes a series of sequential operations, starting from selection of strains resistant to elevated temperatures, their characterization, preparation of inoculum optimization of fermentation technology, product concentration, mixing and defining appropriate drying procedure with stabilizing substances [5–7].

Application of balanced feed rations supplemented with proiotics enables to raise stock breeding efficiency by ruling out outbreaks of acute and chroinic gastrointestinal pathologies in farm animals.

Aim of this chapter was elaboration of complex bacterial probiotic to facilitate biological accessibility of fodder, to promote immune correction and activation of metabolic processes in reared swine and poultry.

12.2   MATERIALS AND METHODS

Bacterial strains *Bacillus subtilis* BIM B 497 and *B. subtilis* BIM B 713 showing antimicrobial activity against strictly and facultatively pathogenic microorganisms and hydrolase (protease, cellulase, xylanase) enzyme activity were engaged in this study.

The cultures are deposited at Belarusian collection of nonpathogenic microorganisms.

Pathogenic bacteria *Escherichia coli, Staphylococcus* sp. isolated by our team and kindly provided by colleagues from Grodno State Agrarian University and S.V. Vyshelessky Institute of Experimental Veterinary Research were used as test objects. Submerged fermentation of bacteria *B. subtilis* BIM B 497 *B. subtilis* BIM B 713 was carried out at temperature 30°C during 3–4 days in Erlenmeyer flasks on the shaker (180–200 rpm), in 10 and 100 l fermentors (aeration rate 1 L air/1 medium · min, agitation rate 200–220 rpm) on nutrient media with molasses as a carbon source. Vegetative seed material pregrown for 1–2 days was applied at the ratio 10% v/v for inoculation of nutrient medium.

To optimize fermentation conditions of *B. subtilis* BIM B 497 and *B. subtilis* BIM B 713 in the bioreactor of 100 l capacity aeration rate was varied from 0.7 to 1.2 L air/l·min at agitation rate 200–220 rpm. For each regime growth parameters were calculated and dynamics of antagonistic activity was examined. Microbiological and biochemical control of fermentation process and accumulation of antimicrobial metabolites was conducted by taking and subsequent analysis of broth samples, while sporulation phenomenon was monitored microscopically. Cell and spore titers of bacteria were determined using finite dilutions technique [8]. Activity of antagonistic bacteria was evaluated by wells method [9] via diameter of zones indicating inhibition of growth in colonies of pathogenic test cultures.

The content of free reducing substances was calculated as previously described [10]. For estimation of total sugars preliminary acid hydrolysis of the medium was performed.

Hydrolase activity of sporulating bacteria was assessed using qualitative and quantitative methods.

Proteolytic activity of examined cultures was assayed on gelatin and casein as the substrates [11]. Quantitatively protease activity of *B. subtilis* BIM B 497 and *B. subtilis* BIM B 713 was measured by Anson method modified by Petrova and Vintsunaite envisaging enzymatic hydrolysis of sodium caseinate. One unit of proteolytic activity was defined as the amount of enzyme catalyzing release of protein substrate hydrolysis products not precipitated by trichloroacetic acid to the extent corresponding to 0.01 OD_{280} rise of reaction mixture at 37 °C during 1 min. Protease activity was expressed in units per 1 mL of cultural liquid (U/mL) [12].

Production and activity of cellulases was analyzed via their impact on specific substrates: β-glucanases on Na-CMC, β-glucosidases on cellobiose.

Diameter of clarification zones around grown colonies after staining Petri plates with Congo red dye indicated activity of generated enzymes [13].

Cellulase activity of the strains selected at initial research stage was estimated as the ability to hydrolyze soluble carboxyl methyl cellulose of medium viscosity and to produce reducing sugars determined with the aid of 3,5-dinitrosalicylic acid and preplotted glucose calibration curve [14].

Amount of enzyme sufficient to catalyze Na–CMC hydrolysis and yield 1 μmole of reducing sugars (calculated as glucose) during 1 min under experimental conditions (temperature 50°C, pH 4.7, duration of the process 10 min) was assumed to be equivalent to 1 unit of cellulase activity (U/mL).

Cellulolytic activity of C_x-enzyme was assayed as its capacity to reduce viscosity of Na-CMC solution using Ostwald viscosimeter at 28°C [15].

Xylanase activity of the cultures was evaluated via their impact on xylan serving as the substrate. The effect was visually observed as the diameter of clear zones encircling grown colonies after staining of Petry plates with Congo red [13].

Quantitatively xylanase activity was determined by the ability of microbial strains to hydrolyze xylan and generate reducing sugars assayed with 3,5-dinitrosalicylic acid and xylose calibration curve [16]. 1 unit of xylanase activity was defined as the amount of enzyme essential to release 1 μmole of glucose from xylan substrate in 1 min under standard conditions.

Veterinary-toxicological trials of probiotic and experiments testing efficiency of test batch in vitro and in vivo were performed by researchers of Grodno State Agrarian University.

Studies on efficiency of probiotic were arranged at lab of Grodno State Agrarian University and at swine-breeding farm Meshetniky, Grodno region. The piglets aged 50–55 days (postsuckers) were divided into 2 groups: experimental (57 animals) and control (54 heads).

Effect of probiotic test batch was analyzed at standard farm facilities in compliance with the established feeding and upkeep technologies, veterinary schemes.

Test group ration was supplemented with probiotic at the dose 1.0–1.5 kg per 1 ton of composite fodder. The experiment lasted 45 days.

Clinical monitoring, control of piglet growth and development was carried out throughout the whole observation period. Effect was checked taking into account parameters of productivity (live weight, average daily and relative weight gains), feed conversion, fodder expenses per 1 kg of live body increment.

The obtained results were analyzed using Microsoft Excel software. Statistical processing of experimental data envisaged calculation of arithmetical means and their confidential ranges for 95% probability level [17].

12.3   RESULTS AND DISCUSSION

Performed investigations resulted in isolation of sporulating bacteria with high antagonistic activity towards animal microbial pathogens and elevated enzyme activities (protease, cellulase, xylanase) (Table 12.1; Fig. 12.1).

TABLE 12.1   Antagonistic Activity of Selected Bacterial Isolates

Isolate	Diameter of test culture growth inhibition zone, mm		
	E. coli 39A	*E. coli* K3	*Staphylococcus* sp.
Сп 36	22.0±0.2	23.5±0.3	28.0±0.3
Кл 53	**25.5±0.2**	**26.0±0.4**	**32.0±0.4**
130	**25.0±0.2**	**26,5±0,2**	**31.5±0.3**
133	25.0±0.2	26.5±0.2	30.0±0.3
146	**23.5±0.2**	**27.5±0.3**	**32.5±0.3**
Сп 54	24.0±0.6	25.0±0.4	29.0±0.5
67	**27.0±0.2**	**25.0±0.6**	**33.0±0.9**
Сп 46	26.0±0.9	23.0±0.4	29.0±0.8
355	**26.0±0.7**	**27.0±0.5**	**31.0±0.3**
359	25.0±0.6	28.0±0.3	30.0±0.5

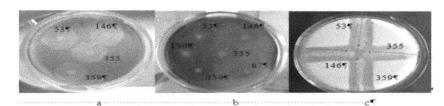

FIGURE 12.1   Zones of Na-CMC (a), xylan (b), casein (c) hydrolysis by selected bacterial isolates.

Comparative appraisement of antibacterial activity of our most active isolates versus the known collection strains-antagonists from genus *Bacillus* displaying antimicrobial action against tested pathogenic species has revealed that newly selected isolates are distinguished by higher suppressing potential in respect of bacteria *E. coli* 39 A, *E. coli* K3. However, regarding *Staphylococcus* sp. the maximal activity was recorded for collection culture *B. subtilis* BIM B 497 (Table 12.2).

TABLE. 12.2 Comparative Appraisal of Antagonistic Activity in New Isolates and Collection Strains of Genus *Bacillus*

Antagonistic bacteria	Diameter of test culture growth inhibition zone, mm		
	E. coli 39A	*E. coli* K3	*Staphylococcus* sp.
B. subtilis BIM B-454 (active principle of probiotic Bacinil)	21.5±0.7	23.0±0.2	30.5±0.3
B. subtilis BIM B-497 (key ingredient of Vetosporin)	23.0±0.5	27.5±0.3	35.5±0.3
B. pumilus BIM B-263 (Enatin basic element)	26.0±0.3	22.0±0.2	31.0±0.3
Kl 53	**26.0±0.2**	**25.0±0.4**	**31.5±0.4**
130	25.5±0.2	26.5±0.3	31.0±0.3
67	25.5±0.2	23.0±0.5	33.0±0.8
355	26.0±0.7	26.5±0.5	31.0±0.3
359	25.0±0.6	28.0±0.3	30.0±0.5

Vivid interest as constituents of novel probiotic was aroused by two cultures with mutually complementary properties: isolate Kl 53 possessing superior enzymatic and antimicrobial activities over the other variants, and collection entry *B. subtilis* BIM B-497 showing increased antagonistic activity (major component of probiotic Vetosporin).

In accordance with morphological-cultural and physiological-biochemical characteristics strain Kl 53 was identified as *Bacillus subtilis* and deposited at collection of nonpathogenic microorganisms, Institute of Microbiology, National Academy of Sciences, Belarus under acronym BIM B 713.

Certificate stating that is not pathogenic and not toxigenic for warm-blooded animals and may be recommended for use in microbiological industrial processes was granted.

Compatibility tests were conducted for the most activate bacterial cultures. It was found that selected *B. subtilis* strains lack cross antagonistic activity as the argument in favor of joining them into microbial association.

For formulation of probiotic based on cultures *B. subtilis* BIM B-497 and *B. subtilis* BIM B-713 the optimal ratio of constituents was chosen using antagonistic activity as a selection criterion. It was demonstrated (Fig. 12.2) that the top value of antagonistic activity among examined mixed cultures differing in cell titer proportions, CFU/mL 1:1, 1:2, 2:1 was registered in the variant combining bacilli in equal percentage 1:1.

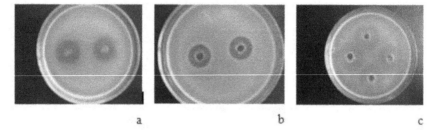

a   b   c

FIGURE 12.2   Diameter of growth inhibition zone for test culture of facultatively pathogenic strain *E. coli* 39A caused by mono (a – *B. subtilis* BIM B-497, b –*B. subtilis* BIM B-713) and binary (c –*B. subtilis* BIM B-497 plus *B. subtilis* BIM B-713 – 1:1 ratio) cultures of antagonistic bacteria.

Process of probiotic production aimed at upgrading fodder manufacturing technology envisages submerged fermentation of sporulating bacteria *B. subtilis* BIM B-497 and *B. subtilis* BIM B-713 ensuring comprehensive control of strain growth and development. The final stage of fabrication comprises preparing mixture of bacterial cultural liquids, adding the binder (wheat flour) to the mix in 1:2 solids ratio, spray drying of bio-preparation (input temperature – 160°C, output temperature - 70°C) to 5% moisture content, packaging and labeling of the end product.

Experiments conducted to seek sources of carbon and nitrogen essential for balanced growth and development of bacteria and synthesis of antimicrobial metabolites defined the following optimal composition of nutrient media for fermentation of *B. subtilis* BIM B-497 and *B. subtilis* BIM B-713 (g/L): molasses – 30.0, $K_2HPO_4 \cdot 3H_2O$–7.0; KH_2PO_4–3.0; $MgSO_4 \times 7H_2O$–0.1; $(NH_4)_2SO_4$–1.5; $Na_3C_6H_5O_7 \times 3H_2O$ – 0.5; corn steep liquor–2.5; tap water – up to 1 L [18].

Studies on optimization of *B. subtilis* BIM B-497 and *B. subtilis* BIM B-713 cultural conditions indicated that maximal antagonistic activity was achieved in temperature range 30–37°C [18], (Table 12.3).

TABLE. 12.3 Effect of Temperature on Growth and Antagonistic Activity of *B. subtilis* BIM B–713

Temperature, °C	Fermentation time, h	Titer		Diameter of growth inhibition zone of test culture *E. coli* 39A, mm
		CFU/mL	Spores/mL	
21	24	3.1×10^6	2.3×10^4	10.0 ± 0.3
	48	4.2×10^6	3.3×10^5	13.5 ± 0.3
	72	1.7×10^7	1.3×10^6	16.0 ± 0.3
30	24	1.5×10^8	6.4×10^6	21.5 ± 0.5
	48	2.1×10^9	5.2×10^8	29.0 ± 0.7
	72	2.5×10^9	8.4×10^8	31.5 ± 0.4
34	24	2.0×10^8	4.6×10^8	22.0 ± 0.4
	48	7.2×10^8	5.4×10^8	26.0 ± 0.6
	72	2.0×10^9	6.0×10^8	28.5 ± 0.7
37	24	2.3×10^9	2.0×10^8	21.0 ± 0.5
	48	2.4×10^9	2.0×10^8	23.5 ± 0.3
	72	2.5×10^9	2.0×10^8	25.0 ± 0.5

This temperature interval was also most favorable for growth. Biomass concentration reached 3.5–4.0 g/L at almost complete utilization of sugars in the medium (over 90%). Decrease of temperature down to 21°C resulted in deceleration of growth rate and reduced accumulation of metabolites with antimicrobial activity.

A vital feature of examined strains is the ability to grow in a broad pH spectrum 4.0–9.0. The best conditions for expression of antagonistic activity were attained at initial active acidity of the medium 6.0–8.0 and pH optimum 7.0.

Beyond the afore-mentioned pH scope antimicrobial activity of *B. subtilis* BIM B-497 and *B. subtilis* BIM B-713 tended to decline.

A crucial factor for acceleration of growth and production of antimicrobial metabolites during submerged culture is supply of sufficient oxygen amount.

It was observed that low oxygen flow rate (0.7 L/L·min) adversely affected growth characteristics and antagonistic properties of strain *B. subtilis* BIM B-713.

Intensification of aeration to 1.0–1.2 L/L·min leads to more complete consumption of nutrient substrate and, as a consequence, rise in total number of cells and spores in cultural medium.

Maximal antagonistic activity with respect to *E. coli* 39A was shown, growth inhibition zones of test cultures equaled 30.0–31.5 mm.

Growth parameters, antagonistic and enzymatic activities of tested bacteria are similar under such conditions, yet enhanced pumping of air inevitably provokes severe foam generation coinciding in time with active growth of the culture which makes control of fermentation process more complicated (Table 12.4).

Series of experiments modeling various aeration regimes during fermentation of *B. subtilis* BIM B-497 were completed earlier [18].

Summing up, optimal conditions for growth and accumulation of antimicrobial metabolites were set for cultures *B. subtilis* BIM B-497 and *B. subtilis* BIM B-713 grown for 48–52 h in lab fermentor at pH 6.9–7.2, temperature 34±2°C (*B. subtilis* BIM B-497), 30±2°C (*B. subtilis* BIM B-713), agitation rate 200–220 rpm, aeration rate 1 L/L min in the medium of the following composition, g/L: molasses – 30.0, $K_2HPO_4·3H_2O$–7.0; KH_2PO_4–3.0; $MgSO_4×7H_2O$–0.1; $(NH_4)_2SO_4$–1.5; $Na_3C_6H_5O_7×3H_2O$ – 0.5; corn steep liquor–2.5; tap water to 1 L.

Dry form of probiotic was derived by mixing cultural liquids of *B. subtilis* BIM B–497 and *B. subtilis* BIM B–713 in 1:1 ratio of cell titers (CFU/mL) with subsequent supply of the binder (wheat flour in proportion of solids 1:2).

Dehydration of the preparation was carried out by feeding its components after preliminary gentle stirring into spray drying unit.

In physical-chemical, microbiological parameters and flavor probiotic meets the criteria and standards specified in Table 12.5.

TABLE 12.4 Dynamics of Growth and Antimicrobial Activity of *B. Subtilis* BIM B-713 during Submerged Fermentation in Laboratory Fermentor ANKUM 2 M

Duration of fermentation, h	pH	Sugar contents, g/L		Diameter of growth inhibition zones, mm	Titer	CFU/mL	Spores/mL	Proteolytic activity, u/mL
		Free reducing substances	Inverted reducing substances	*E. coli* 39A	*Staphylococcus* sp.			
0	6.69	7.25	12.60	18.0	20.0	1.1×10^7	-	-
12	6.50	4.20	8.80	19.5	22.0	7.8×10^7	5.2×10^5	13.67
24	6.85	5.9	6.00	30.3	32.0	2.3×10^8	4.5×10^8	14.3
36	7.05	5.10	5.90	31.0	35.0	2.8×10^9	5.2×10^8	12.77
48	6.90	4.70	6.0	31.5	36.0	2.9×10^9	1.0×10^9	19.37
60	6.85	3.00	2.3	26.0	28.0	1.9×10^9	9.5×10^8	20.67
72	6.80	3.25	2.5	24.8	25.0	1.7×10^9	8.0×10^8	15.80

TABLE. 12.5   Main Characteristics of Dry Probiotic

Parameters	Characteristics
Appearance	Powder of light-brown to beige color
Odor	Specific for this type of product
pH	7.1–7.3
Moisture content, %	At least 5%
Titer of viable cells, bln/g	At least 1
Titer of viable spores, bln/g	Minimum 0.5
Antagonistic activity estimated via diameter of growth inhibition zone of *Escherichia coli* 39A, mm	Minimum 18.0

The collected data were used to scale up process of manufacturing new probiotic to pilot-plant level at facilities of Biotechnological Center, Institute of Microbiology, NAS Belarus. Optimization of fermentation parameters in 100 l fermentors proved some correlations revealed during growth of bacterial cultures *B. subtilis* BIM B–497 and *B. subtilis* BIM B–713 in 10 l laboratory fermentors. For instance, it was established that low airflow rate (0.7 L/L·min) had a negative impact on growth characteristics, antagonistic and enzyme activities of strains-producers.

More intense aeration rate (1.0 or 1.2 L/L min) resulted in higher consumption of nutrient substrate and increased total number of microbial cell and spores in the media. It was accompanied by maximal antagonistic activity of the cultures towards test pathogenic species. Aeration rate 1 L/L min was preferential for growth and synthesis of antimicrobial metabolites by *B. subtilis* BIM B–497 and *B. subtilis* BIM B-713 since it allowed to rule out enhanced foam generation in the course of fermentation process.

Temperature regimens applied in lab fermentors for *B. subtilis* BIM B–497 (optimum 34±2°C) and *B. subtilis* BIM B–713 (optimum 30±2°C) were not modified in pilot-plant trials. During fermentation under optimized conditions cell and spore titers, enzyme activity reached their top values by 36–48 h of bacterial growth, whereas peak antagonistic activity against *E. coli* 39A was recorded by 24–48 h (Table 12.6). Prolongation of fermentation resulted in decreased activity and reduced titers of cells

TABLE 12.6 Dynamics of Growth and Antagonistic Activity of Bacteria *B. Subtilis* BIM B-497 and *B. Subtilis* BIM B-713 during Fermentation in Pilot-Plant Bioreactors

Fermentation time, h	pH	Total sugars, g/L	Free reducing substances, g/L	Titer		Diameter of E. coli 39A growth inhibition zone, mm	Proteolytic activity, U/mL
				CFU/mL	Spores/mL		
B. subtilis BIM B-497							
0	7.1	12.2	7.3	1.1×10^7	—	18.0±0.2	—
12	6.8	9.8	6.2	5.8×10^7	5.2×10^5	19.0±0.4	10.67
24	6.9	7.0	5.8	2.3×10^8	4.5×10^8	30.3±0.3	14.3
36	7.1	4.8	5.6	2.5×10^9	5.2×10^8	30.3±0.5	12.77
40	7.0	4.1	5.4	2.8×10^9	5.7×10^8	30.5±0.7	15.80
48	6.9	3.8	4.9	3.2×10^9	1.5×10^9	31.0±0.5	17.37
60	6.9	4.2	3.0	2.0×10^9	9.5×10^8	28.0±0.4	16.67
B. subtilis BIM B-713							
0	7.0	14.0	8.0	1.1×10^6	—	17.0±0.3	—
12	6.8	10.0	5.1	7.8×10^6	4.9×10^5	19.0±0.5	10.4
24	6.8	7.4	4.5	3.7×10^8	4.5×10^7	22.0±0.3	15.3
36	7.0	6.0	4.3	3.7×10^8	7.0×10^8	27.3±0.5	19.2
40	7.2	5.9	4.4	8.1×10^8	1.3×10^9	28.0±0.7	18.8
48	7.3	5.2	4.1	3.8×10^9	1.4×10^9	27.5±0.5	22.7
60	7.2	4.5	3.8	2.9×10^9	1.2×10^9	27.5±0.5	18.8

and spores so that duration of the process over 2 days was regarded as not expedient.

Summing up, it was found that most favorable terms for growth and accumulation of antimicrobial metabolites by bacterial cultures under pilot-plant conditions were achieved during fermentation period 36–48 h at aeration rate 1 L air/l medium min, temperature 30–34°C, agitation rate 200±20 rpm using nutrient media with molasses and ammonium sulfate.

Formulation of probiotic comprised mixing bacterial cultural liquids in equal proportion to cell concentration at least 1×10^9 CFU/ mL, adding wheat flour (solids ratio 1:2), spray-drying the resulting mixture (input temperature 160°C, output temperature 70°C) to 5% moisture content, packaging and labeling of the end product.

Manufactured test batch of probiotic (500 doses) was characterized by the following parameters: appearance – light-brown to beige powder with specific odor; humidity – 5%; pH – 7.2; titer – 1.2×10^{10} CFU/g, 1.0×10^{10} spores/g; antagonistic activity estimated by wells technique via diameter of growth inhibition zones for *E. coli* and *Staphylococcus* sp. – 25.0 and 30.0 mm, respectively; enzyme activities: protease – 17.6 U/mL, xylanase – 3.8 U/mL, cellulase – 1.0 U/mL.

Efficiency trials of probiotic test batch were performed at Meshetniky swine farm, Grodno region by introducing it into the rations of piglets-weanlings aged 50–55 days. The existing feeding technologies and veterinary schemes were not changed.

The experiments demonstrated (Table 12.7) that average daily weight gain in the test group constituted 434 g versus 325 g in the control group, which is 33.6% higher. The relative weight increment was superior to the control by 8.68%.

TABLE 12.7 Average Daily and Relative Increments of Live Body Weight in Weanling Piglets

Parameters	Group		
	Control	Test	% of the control
Average daily weight gain, g	325.0	434.0	133.6
Relative increment, %	56.27	64.95	-

Efficiency of supplying probiotic into the rations of weanlings was also illustrated by such indices as fodder spent per unit of piglet live weight increase and feed conversion.

Our investigations have shown (Table 12.8) that application of probiotic allowed to reduce by 23.6% fodder expenses per 1 kg of body weight rise and feed conversion by 42.8%.

TABLE 12.8   Estimated Fodder Expenses per Unit of Product

Parameters	Control	Test group	% of the control
Average daily weight gain, g	325.0	434.0	133.6
Average diurnal fodder expense per capita, g	1470.0	1500.0	102.0
Fodder spent per 1 kg of body increment, g	4523.1	3456.2	76.4
Feed conversion	13.9	8.0	57.2

12.4   CONCLUSION

In conclusion, performed farm efficiency trials of probiotic test batch fed to piglets as ingredient of standard rations evidenced that this biological product improved tissue nutrition in the body, activated redox and metabolic processes, raised natural resistance and immunobiological reactivity, promoted increase of piglet live weight by 4.7%, average daily weight gain by 33.6%, relative body increment by 8.7% and reduction of feed conversion by 42.8%, fodder expenses per unit of live weight rise by 23.6%. Our novel biopreparation exceeds in biological efficiency all Belarusian analogs and matches recognized foreign probiotics Toyocerin (Germany) and Bacell, Pro-A (Russia) [19, 20].

KEYWORDS

- **Animal pathogens**
- **Bacillus subtilis**
- **Biotechnology**
- **Fermentation**
- **Fodder**
- **Probiotic**
- **Sporulating bacteria**

REFERENCES

1. Malik, N. I., & Panin, A. N. (2009). Proc. Int. Congress, Saint-Petersburg.
2. Ushakova, N. A., et al. (2012). Fundamental nye issledovania, 1, 184.
3. Sverchkova, N. V. et al. (2012). Collected papers, 4, 107 (in Russian).
4. Chikov, A. E., & Pyshmantseva, N. A. (2011). Application of nonconvertional fodder, feed and bioactive supplements in rations of farm stock and poultry. Krasnodar198 p. (in Russian).
5. Leonov, G. V. (2010). www.bti.secna.ru.
6. Scharek, L., et al. (2007). Veterinary Immunology and Immunopathology 120(3–4), 136.
7. Wannaprasat W., et al. (2009). South-east Asian J. Trop Med Public Health, 40(5), 1103.
8. Lysak, V. V., & Zheldakova, R. A. (2002). Microbiology. Methodological recommendations for laboratory studies and tutorial supervision of student work. Belarusian State University, Minsk, 54 p. (in Russian).
9. Segy, J. (1983). Methods of soil microbiology Kolos Press, Moscow, 253 p. (in Russian).
10. Sverchkova, N. V. et al. (2012). Polish Journal of Natural Science, 27(1), 15.
11. Egorov, N. S., et al. (Eds.) (1983). Guidelines for Practical Studies in Microbiology Moscow State University Moscow 215 p. (in Russian).
12. Petrova, I. S., & Vintsunaite, M. M. (1980). Prikl. Boichem. Microbiol 2(2), 322.
13. Osadchaya, A. I., et al. (2009). Microbiological, 5, 41.
14. Methods of assaying cellulase enzyme activity. National standard P. 53046. (2008) (in Russian).
15. Rukhlyadeva, A. P., & Polygalina, G. V. (1981). Methods of evaluating activity of hydrolytic enzymes Lyohkaya i pischev. promyshl, Moscow. 288 p. (in Russian)
16. Methods to determine xylanase enzyme activity. National standard 53047–2008 (in Russian).
17. Tyurin, Ju. N., & Makarov, A. A. (1998). Computerized statistical data processing, INFA-M, Moscow 544 p. (in Russian).
18. Kolomiets, E. I., et al. (2009). Collected papers 2, 231.
19. Zhelamsky, S. V. (2005). Tsenovik, 2, www.tsenovik.ru.
20. Gorkovenko, L. G., et al. (2011). Instructions for applications of probiotics Prolam, Monosporin and Bacell at swine farms Krasnodar, 15 p. (in Russian).

CHAPTER 13

MICROHETEROGENEOUS TITANIUM ZIEGLER-NATTA CATALYSTS: 1,3-DIENE POLYMERIZATION UNDER ULTRASOUND IRRADIATIONS

VADIM P. ZAKHAROV, VADIM Z. MINGALEEV,
IRIVA D. ZAKIROVA, and ELENA M. ZAKHAROVA

CONTENTS

ABSTRACT

Polymerization of butadiene and isoprene under action of microheterogeneous titanium based catalyst with ultrasonic irradiation of the reaction mixture at the initial time is studied. It is shown that ultrasonic irradiation causes transformation multisite catalyst in a quasi-single site. At that activity of dominant site depends on the monomer nature.

13.1   INTRODUCTION

One of the important application areas of ultrasound (US) is catalytic reactions with the participation of low molecular mass compounds and heterogeneous catalysts. The effect of ultrasound on catalytic reactions in the presence of platinum and rhodium catalysts of various dispersities was investigated in Ref. [1]. It was demonstrated that ultrasound can provide for the occurrence of chemical processes that cannot be performed even in the presence of catalysts. It is assumed that the main mechanism of its action on catalytic processes consists in the dispersion of catalyst particles; however, as was shown in Ref. [1], the adhesion of particles can occur during the action of the so-called Bjerknes forces, that is, forces that promote the attraction of particles (primarily small particles) to a deformed bubble followed by their sticking together. As a consequence, the diffusion of reagents to the surface of a particle becomes more pronounced and the rate of the process increases.

It seemed useful to examine the effect of ultra sound on the catalytic polymerization of dienes in the presence of Ziegler–Natta titanium catalysts, because the catalytic system is microheterogeneous and potentially susceptible to the effect of ultrasound. Moreover, although a long time has passed since the development of techniques for the synthesis of stereoregular polydienes, these catalysts or rare-earth compounds with close catalytic mechanisms are in current use for the majority of large tonnage manufacturing of synthetic rubber. The large body of research on the synthesis of steroregular polydienes and the absence of substantial changes in the organization of their commercial production are due to the lack of understanding of some specific features of Ziegler catalysis: in particular, the origin of the multisite nature and the relationship of the reactivity distribution of active sites to the stereoregulating activity, to the micro heterogeneity of a catalyst, to the kinetic parameters of the process, etc. Thus, US irradiation, which can simultaneously affect several parameters of the process, is a useful tool for the study of specific features of polydiene synthesis with Ziegler–Natta catalysts.

The aim of this study is to investigate the polymerization of butadiene and isoprene in the presence of the catalytic system $TiCl_4$-$Al(iso$-$C_4H_9)_3$ under US irradiation of the reaction mixture during its formation.

13.2 EXPERIMENTAL PART

The reaction mixture was exposed to US irradiation on a UZDN–2T apparatus equipped with a conical irradiator operating at a frequency of 22 kHz; the maximum power was 400 W at a current strength of 25 mA. A 500-cm^3-reaction flask was hermetically connected to an ultrasound irradiator to prevent contact of the reaction mixture with the atmosphere.

The titanium catalytic complex was prepared through pouring of toluene solutions of TiCl4 and Al(iso-C4H9)3 into a separate reaction vessel. Then, the catalytic complex was aged at 0°C under stirring for 30 min to achieve the maximum activity. The optimum ratio of the catalytic system components, Al/Ti, was dependent on the nature of the monomer; these ratios were 1.4 and 1.1 for butadiene and isoprene, respectively.

Polymerization was performed in toluene free of trace moisture and admixtures that deactivate the catalyst. The temperature of polymerization was 25($\pm$1)°C. The catalyst concentration was 5 mmol/L, and the monomer concentration was 1.5 mol/L. These values correspond to the maximum activities and stereo specificities of the catalyst for these monomers. Polymerization was conducted with the use of two methods.

The preliminarily prepared titanium catalyst was added to the monomer solution in a flow of argon. This time was taken as the onset of polymerization. The process was performed under constant stirring with a magnetic stirrer. The synthesis was conducted in a manner similar to that described above, but at the time of catalyst addition, the reaction mixture was subjected to ultrasonic stirring during the first minute of polymerization. Published data show that US irradiation can initiate the polymerization of some monomers in the absence of initiating agents. To estimate the contribution of ultrasound-initiated polymerization, US irradiation of toluene solutions of the monomers (butadiene and isoprene) was performed for 1 min in the absence of the titanium catalytic complex. Then, polymerization was performed under constant stirring with a magnetic stirrer. The polymer was sampled directly from the reaction mixture in a flow of argon. To stop polymerization at the predetermined time of synthesis, a calculated amount of methanol, which facilitates decomposition of the catalytic complex and precipitation of the polymer from toluene solutions, was added to the reaction mixture. The residual catalyst was removed from the polymer via repeated washing of the sample with methanol (pH ~ 5–6). At the final stage, the polymer was reprecipitated and the samples were dried in vacuum to a constant weight. The yields of the polymers were determined gravimetrically.

The molecular-mass characteristics of the polymers were determined via GPC on a Waters GPC-2000 instrument. Calibration was made relative to narrowly dispersed polystyrene standards.

The microstructure of polybutadiene was studied via IR spectroscopy on a Shimadzu IR Prestige spectrometer. Analysis was performed with the use of polymer films applied on KBr glasses. The films were cast from toluene solutions. The microstructure of polyisoprene was studied via ^{1}H NMR spectroscopy on a Bruker AM–300 spectrometer. Deuterochloroform was used as a solvent.

13.3   RESULT AND DISCUSSION

When the toluene solutions of isoprene and butadiene were U.S. irradiated without any catalyst, ultrasound did not initiate polymerization processes under the given experimental conditions. After irradiation of monomer solutions and further stirring with a magnetic stirrer, even trace amounts of the polymers were not detected, as demonstrated by the absence of characteristic turbidity of solution upon the addition of methanol.

In the absence of ultrasonic irradiation of the reaction mixture (method 1), polybutadiene with a wide polymodal molecular-mass distribution is formed at initial conversions; in contrast, polyisoprene is characterized by a narrower unimodal molecular-mass distribution (Fig. 13.1). During ultrasound treatment, the molecular-mass distribution of polybutadiene formed at initial conversions becomes significantly narrower owing to reduction in the shares of both low and high molecular mass fractions (Fig. 13.1). Under similar conditions, the width of the molecular-mass distribution of polyisoprene decreases slightly (Fig. 13.1); in this case Mw of the polymer tend to increase (Fig. 13.2). Thus, regardless of the nature of a monomer, US irradiation facilitates the synthesis of polybutadiene and polyisoprene with narrower molecular-mass distributions. Since the statistical width of the molecular-mass distributions for polybutadiene and polyisoprene synthesized via methods 1 and 2 is greater than the poly dispersity index for the most probable Flory distribution, it is reasonable to suggest that the titanium catalytic system is kinetically heterogeneous.

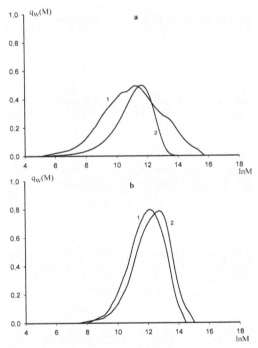

FIGURE 13.1   Molecular-mass distributions of (a) polybutadiene and (b) polyisoprene: (*1*) method 1 and (*2*) method 2. Conversion is 1–3%.

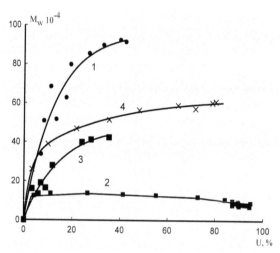

FIGURE 13.2 Weight-average molecular masses of (*1, 2*) polybutadiene and (*3, 4*) polyisoprene vs. conversion: (*1, 3*) method 1 and (*2, 4*) method 2.

In terms of the multisite nature of Ziegler–Natta catalytic systems [2, 3], each functioning type of active site generates polymer fractions with a certain molecular mass and stereoregular composition; that is, polymerization sites are kinetically and stereospecifically heterogeneous.

On the basis of experimental molecular-mass distributions of the polymers, the inverse task was solved via Tikhonov's regularization method [3]; as a result, the functions of distribution of active sites over the probability of chain termination were obtained (Figs. 13.3 and 13.4). The titanium catalytic system is characterized by a polymodal distribution function, where each maximum corresponds to a certain type of polymerization sites. Under our conditions of polybutadiene synthesis, there are four types of active sites that form macromolecules with the following most probable molecular masses: $\ln M = 9.2$–10.4 (I), 11.2–11.4 (II), 12.9–13.2 (III), and 14.1–14.7 (IV) (Fig. 13.3). The position of maxima on the curves is conversion independent. This implies that each type of site forms macromolecules of a certain length throughout the process.

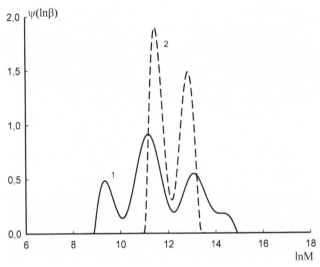

FIGURE 13.3   Kinetic-heterogeneity distributions of active sites for polymerization of butadiene: (*1*) method 1 and (*2*) method 2. Conversion is 1–3%.

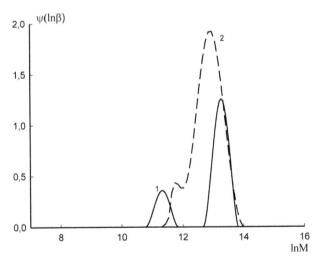

FIGURE 13.4   Kinetic heterogeneity distributions of active sites for polymerization of isoprene: (*1*) method 1 and (*2*) method 2. Conversion is 1–3%.

When ultrasound treatment was used at the onset of polymerization of butadiene, the titanium catalytic system featured only type II and type III polymerization sites, as evidenced by the occurrence of only two maxima on the curve of the distribution of active sites over kinetic heterogeneity (Fig. 13.3). A decrease in the types of sites during polymerization via method 2 is responsible for narrowing of the molecular-mass distribution of polybutadiene.

During the synthesis of polybutadiene via method 1, the activities of type I and type II sites, which generate polybutadiene fractions with relatively low molecular masses, decrease during polymerization (Fig. 13.5). In contrast, the activities of type-III and type IV sites (Fig. 13.5) that form a high molecular mass polydiene increase with monomer conversion.

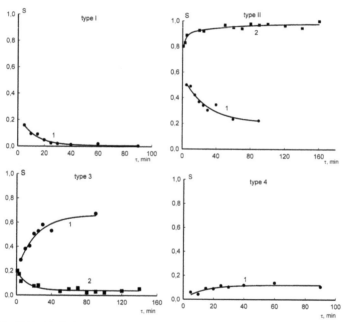

FIGURE 13.5   Variation in the kinetic active sites of butadiene polymerization: (*1*) method 1 and (*2*) method 2.

UV irradiation brings about an abnormal change in the activity of functioning types of polymerization sites. The activity of type II polymerization sites tends toward unity (Fig. 13.5); that is, the multisite titanium catalytic system is transformed into a quasi-single site system during ultrasonic treatment. Thus already at 3 min of polymerization, the fraction of the monomer polymerized on propagating type II sites achieves 90%.

When the time of polymerization is increased to 60 min, this fraction of the monomer becomes as high as 98%.

Solution of the inverse problem of formation of the molecular-mass distribution for the polymerization of isoprene showed that, when polymerization was performed via method 1, the titanium catalytic system was characterized by the presence of type II and type III active sites (Fig. 13.4).

In this case, active sites forming high-molecular mass polyisoprene show higher activities than those in the polymerization of butadiene. The kinetic activity of type II sites decreases during the polymerization of isoprene under US irradiation (Fig. 13.6). In this case, the activity of type III sites increases significantly (Fig. 13.7). Thus, at a time of polymerization of 50 min, the fraction of isoprene polymerized on type III sites is as high as 95%.

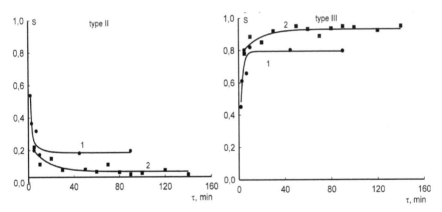

FIGURE 13.6   Variation in the kinetic active sties of isoprene polymerization: (*1*) method 1 and (*2*) method 2.

After US irradiation of the reaction mixture, the fraction of cis-1,4-units in polybutadiene decreases, while the fraction of trans1,4-units increases with an increase in conversion (Fig. 13.7). Polybutadiene formed at high monomer conversions contains equal amounts of cis and trans units (48.8% trans1,4–48.8% cis-1,4-, and 2.4% 1,2-units). US treatment has no effect on the content of 1,2-units in polybutadiene. At the same time, ultrasound mixing does not alter the cis stereospecificity of the titanium catalytic system in the polymerization of isoprene.

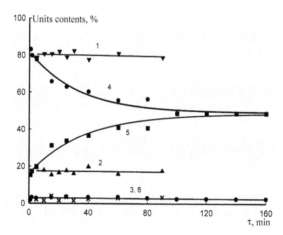

FIGURE 13.7 Contents of (*1, 4*) cis-1,4-, (*2, 5*) trans1,4-, and (*3, 6*) 1,2-units in polybutadiene: (*1–3*) method 1 and (*4–6*) method 2.

An examination of dispersity of the microheterogeneous titanium catalyst in the absence of ultrasound showed that, for catalyst particles, the most probable weight-average radius is 4.2 μm (Fig. 13.8). In this case, the share of fractions of relatively large particles (6–12 μm) decreases. Thus, the ultrasound irradiation of the catalytic system leads to narrowing of the size distribution of particles without any noticeable change in the most probable radius. This effect is most probably associated with the fact that, during US mixing, the size distribution of particles is the result of tow processes: on the one hand, the dispersion of large catalyst particles during the action of ultrasound vibrations of the solid medium and, on the other hand, the adhesion of small catalyst particles during the action of Bjerknes forces.

The kinetic curves for the polymerization of butadiene and isoprene with the titanium catalytic system in the absence of US irradiation (method 1) are almost coincident (Fig. 13.9). US irradiation (method 2) brings about an increase in the initial rates of butadiene and isoprene polymerization and accelerates accumulation of the polymer in the system. In this case, the initial rates of polymerization of butadiene and isoprene increase owing to an increase in the rate constant of chain propagation without any marked changes in the concentration of active sites. This result correlates with the estimation of dispersity of the catalytic system, specifically, with the absence of changes in the most probable size of catalyst particles. In the case of polymerization of butadiene via method 2, the numerical values of rate constants of chain termination increase.

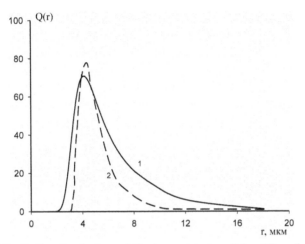

FIGURE 13.8 Radius distributions of titanium catalyst particles: (*1*) method 1 and (*2*) method 2.

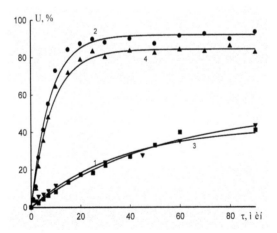

FIGURE 13.9 Yields of (*1, 2*) polybutadiene and (*3, 4*) polyisoprene vs. time of polymerization: (*1, 3*) method 1 and (*2, 4*) method 2.

13.4 CONCLUSION

The ultrasonic irradiation of the reaction mixture at the initial time of the polymerization of butadiene and isoprene under action Ti-Al catalyst result in an increase of process rate. Polymerization occurs at the single active site reactivity of which depends on the nature of diene. With ultrasonic treatment the reactivity of the active sites of polymerization of isoprene greater than for the butadiene polymerization. Reduction of polymerization reactivity site butadiene reduces cis-specificity titanium catalyst. In all cases ultrasound irradiation produces a polydienes with a narrow molecular weight distribution.

ACKNOWLEDGMENTS

This study was financially supported by the Council of the President of the Russian Federation for Young Scientists and Leading Scientific Schools Supporting Grants (project no. MD-4973.2014.8).

KEYWORDS

- **Diene Polymerization**
- **Heterogeneous Catalysis**
- **Titanium Ziegler-Natta Catalyst**
- **Ultrasound**

REFERENCES

1. Margulis, M. A. (1986). *Sonochemical Reactions and Sonoluminescence,* Moscow: Khimiya, 286 p.
2. Yu. V. (2012). Kissin, *Journal of Catalysis* 292, 188–200.
3. Monakov, Y. B., Sigaeva, N. N., & Urazbaev, V. N. (2005). *"Active Sites of Polymerization" Multiplicity: Stereospecific and Kinetic Heterogeneity,* Leiden: Brill Academic, 397 p.

CHAPTER 14

MICROHETEROGENEOUS TITANIUM ZIEGLER-NATTA CATALYSTS: THE INFLUENCE OF PARTICLE SIZE ON THE ISOPRENE POLYMERIZATION

ELENA M. ZAKHAROVA, VADIM Z. MINGALEEV, and
VADIM P. ZAKHAROV

CONTENTS

ABSTRACT

The effect particle size of microheterogenous catalyst $TiCl_4-Al(iso-C_4H_9)_3$ on the basic patterns of isoprene polymerization is studied. Fraction of particles with a certain size isolated from the mixture of particles, which is formed by reacting the initial components of the catalyst and its following exposition. The most active in isoprene polymerization is particles fraction with starting diameter of 0.7–4.5 µm. In these particles preferably localized highly active polymerization site. Doping diphenyloxide and piperylene, lower exposition temperature, hydrodynamic actions in turbulent flows result in formation of nearly mono disperse catalyst with diameter of 0.15–0.18 µm. In this case there is a shift in activity spectrum of polymerization sites towards formation of single site with high activity. **PACS:** 82.65+r, 82.35–x.

14.1   INTRODUCTION

The formation of highly stereo regular polymers under the action of micro heterogeneous Ziegler–Natta catalysts is accompanied by broadening of the polymer MWD [1, 2]. This phenomenon is related to the kinetic heterogeneity of active sites (AC) [1, 3, 4]. The possible existence of several kinetically nonequivalent AC of polymerization correlates with the nonuniform particle size distribution of a catalyst [4]. At present time much attention is given to study the influence of micro heterogeneous catalysts particle size on the properties of polymers [5, 6]. However, almost no detailed study of the effect particle size catalysts on their kinetic heterogeneity.

The micro heterogeneous catalytic system based on $TiCl_4$ and $Al(iso-C_4H_9)_3$ that is widely used for the production of the cis-1,4-isoprene. Our study has shown [7] that the targeted change of the solid phase particle size during the use of a tubular turbulent reactor at the stage of catalyst exposure for many hours is an effective method for controlling the polymerization process and some polymer characteristics of isoprene. We suppose that the key factor is the interrelation between the reactivity of isoprene polymerization site and the size of catalyst particles on which they localize.

The aim of this study was to investigate the interrelation between the particle size of a titanium catalyst and its kinetic heterogeneity in the polymerization of isoprene.

14.2 EXPERIMENTAL PART

Titanium catalytic systems (Table 14.1) were prepared through two methods.

14.2.1 METHOD 1

At 0 or −10°C in a sealed reactor 30–50 mL in volume with a calculated content of toluene, calculated amounts of $TiCl_4$ and $Al(iso\text{-}C_4H_9)_3$ toluene solutions (cooled to the same temperature) were mixed. The molar ratio of the components of the catalyst corresponded to its maximum activity in isoprene polymerization. The resulting catalyst was kept at a given temperature (Table 14.1) for 30 min under constant stirring.

14.2.2 METHOD 2

After preparation and exposure of titanium catalysts via method 1, the system was subjected to a hydrodynamic action via single circulation with solvent through a six-section tubular turbulent unit of the diffuser-confusor design [8] for 2–3 s.

The catalyst was fractionated through sedimentation in a gravitational field. For this purpose calculated volumes of catalysts prepared through methods 1 and 2 were placed into a sealed cylindrical vessel filled with toluene. In the course of sedimentation, the samples were taken from the suspension column at different heights, a procedure that allowed the separation of fractions varying in particle size.

The titanium concentrations in the catalyst fractions were determined via FEK colorimeter with a blue light filter in a cell with a 50 mm thick absorbing layer. A K_2TiF_6 solution containing 1×10^{-4} g Ti/mL was used as a standard.

The catalyst particle size distribution was measured via the method of laser diffraction on a Shimadzu Sald-7101 instrument.

TABLE 14.1 Titanium Catalytic Systems and Their Fractions Used for Isoprene Polymerization

Catalyst	Labels	Molar ratio of catalyst components			T, °C	Method	Range of particle diameters in fractions of titanium catalysts, μm		
		Al/Ti	DPO/Ti	PP/Al			Fraction I	Fraction II	Fraction III
$TiCl_4$–$Al(i\text{-}C_4H_9)_3$	C-1	1	-	-	0	1	0.7–4.5	0.15–0.65	0.03–0.12
						2	–	0.20–0.7	0.03–0.18
$TiCl_4$–$Al(i\text{-}C_4H_9)_3$ –DPO	C-2	1	0.15	-	0	1	0.7–4.5	0.15–0.65	0.03–0.12
						2	–	0.15–0.68	0.03–0.12
$TiCl_4$–$Al(i\text{-}C_4H_9)_3$ –DPO–PP	C-3	1	0.15	0.15	0	1	–	0.12–0.85	0.03–0.10
						2	–	0.15–0.80	0.03–0.12
$TiCl_4$–$Al(i\text{-}C_4H_9)_3$ –DPO–PP	C-4	1	0.15	0.15	-10	1	–	0.12–0.45	0.03–0.10
						2	–	0.12–0.18	0.04–0.11
Averaged ranges, μm							0.7–4.5	0.15–0.69	0.03–0.14

Note: DPO – diphenyloxide, PP – piperylene.

Before polymerization, isoprene was distilled under a flow of argon in the presence of Al(iso-C$_4$H$_9$)$_3$ and then distilled over a TiCl$_4$–Al(iso-C$_4$H$_9$)$_3$ catalytic system, which provided a monomer conversion of 5–7%. The polymerization on fractions of the titanium catalyst was conducted in toluene at 25 °C under constant stirring. In this case, the calculated amounts of solvent, monomer, and catalyst were successively placed into a sealed ampoule 10–12 mL in volume. The monomer and catalyst concentrations were 1.5 and 5×10^{-3} mol/L, respectively. The polymerization was terminated via the addition of methanol with 1% ionol and 1% HCl to the reaction mixture. The polymer was repeatedly washed with pure methanol and dried to a constant weight. The yield was estimated gravimetrically.

The MWD of poly isoprene was analyzed via GPC on a Waters GPC-2000 chromatograph equipped with three columns filled with a Waters microgel (a pore size of 103–106 A) at 80°C with toluene as an eluent. The columns were preliminarily calibrated relative to Waters PS standards with a narrow MWD (M$_w$/M$_n$=1.01). Analyzes were conducted on a chromatograph, which allows calculations with allowance for chromatogram blurring. Hence, the need for additional correction of chromatograms was eliminated.

The microstructure of poly isoprene was determined via high-resolution 1H NMR spectroscopy on a Bruker AM–300 spectrometer (300 MHz).

The MWD of cis-1,4-polyisoprene obtained under the aforementioned experimental conditions, q$_w$(M), were considered through the equation.

$$q_w(M) = \int_0^\infty \psi(\beta) M \beta^2 \exp(-M\beta) d\beta \qquad (1)$$

where β is the probability of chain termination and ψ(β) is the distribution of active site over kinetic heterogeneity, M is current molecular weight.

As shown in Ref. [9], Eq. (1) is reduced to the Fredholm integral equation of the first kind, which yields function ψ(β) after solution via the Tikhonov regularization method. This inverse problem was solved on the basis of an algorithm from Ref. [9]. As a result, the function of the distribution over kinetic heterogeneity in ψ(lnβ)–lnM coordinates with each maximum related to the functioning of AC of one type was obtained.

14.3   RESULTS

After mixing of the components of the titanium catalyst, depending on its formation conditions, particles 4.5 μm to 30 nm in diameter, which are separated into three arbitrary fractions, are formed (Table 14.1).

During the formation of catalyst C-1 via method 1, the fraction composed of relatively coarse particles, fraction I, constitutes up to 85% (Fig. 14.1). In method 2, the hydrodynamic action on the titanium catalyst formed under similar conditions results in an increase in the content of fraction II. Analogous trends are typical of catalyst C-2. The catalyst modification with piperylene additives, catalyst C-3, is accompanied by the disappearance of fraction I and an increase in the content of fraction II (Fig. 14.1), as was found during the hydrodynamic action on C-1. The hydrodynamic action on a two-component catalyst is equivalent to the addition of piperylene to the catalytic system. The preparation of catalytic complex C-3 via method 2 results in narrowing of the particle size distribution of fraction II owing to disintegration of particles 0.50–0.85 μm in diameter (Fig. 14.1). The reduction of the catalyst exposure temperature to −10°C (catalyst C-4) is accompanied by further disintegration of fraction II (Fig. 14.1). In this case the content of particles 0.19–0.50 μm in diameter decreases to 22% with predominance of particles 0.15–0.18 μm in diameter. The formation of C–4 via method 2 results in additional dispersion of particles of fraction II, with the content of particles 0.15–0.18 μm in diameter attaining 95%.

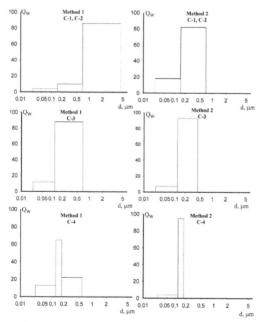

FIGURE 14.1   Fractional compositions of titanium catalyst C-1−C-2 (Table 14.1).

The content of the finest catalyst particles in the range 0.03–0.14 μm, fraction III, attains 5–12% and is practically independent of the catalyst formation conditions. The most considerable changes are shown by particles 0.18–4.50 μm in diameter. Particles of fraction I are easily dispersed as a result of the hydrodynamic action in turbulent flows and the addition of catalytic amounts of piperylene, and their diameter becomes equal to that of particles from fraction II. The decrease in the catalyst exposition temperature from 0 to −10°C with subsequent hydrodynamic action leads to a more significant reduction of particle size and the formation of a narrow fraction.

Isolated catalyst fractions differing in particle size were used for isoprene polymerization. The cis-1,4-polymer was obtained for all fractions, regardless of their formation conditions. The contents of cis-1,4 and 3,4 units were 96–97 and 3–4%, respectively. Coarse particles (fraction I) are most active in isoprene polymerization (method 1) on different fractions of C-1 (Fig. 14.2).

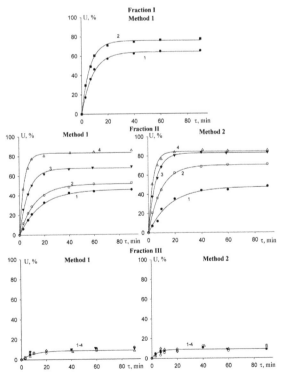

FIGURE 14.2   cis-1,4-Polyisoprene yields U vs. polymerization times τ in the presence of fractions particles of titanium catalysts (1) C-1, (2) C-2, (3) C-3, and (4) C–4.

As the particle size of C–1 decreases, its activity drops significantly. The catalyst modification with diphenyloxide (C-2) has practically no effect on the fractional composition, but the activities of different catalyst fractions change. The most marked increase in activity was observed for fraction I. Catalyst C-3 prepared via method 1 comprises of two fractions, with fraction II having the maximum activity. The decrease of the catalyst exposition temperature to 10°C (C–4) result in further increase in the rate of isoprene polymerization on particles of fraction II.

The hydrodynamic action increases the content of fraction II in C–1, but its activity in isoprene polymerization does not increase (Fig. 14.2). For C–2 the analogous change in the fractional composition is accompanied by an increased activity of fraction II (method 2). The addition of piperylene to C–3 results in a stronger effect on the activity of fraction II under the hydrodynamic action. The change of the hydrodynamic regime in the reaction zone does not affect the activity of fine particles of catalyst fraction III. Isoprene polymerization in the presence of fraction III always has a low rate and a cis-1,4-polyisoprene yield not exceeding 7–12%.

During polymerization with C–1 composed of fraction I (method 1), the weight-average molecular mass of polyisoprene increases with the process time (Table 14.2). Polyisoprene prepared with catalyst fraction II has a lower molecular mass. A more considerable decrease in the weight-average molecular mass is observed during isoprene polymerization on the finest particles of fraction III, with M_w being independent of the polymerization time. The width of the MWD for polyisoprene obtained on fraction I increases with polymerization time from 2.5 to 5.1. The polymer prepared in the presence of catalyst fraction III shows a narrow molecular mass distribution ($M_w/M_n \sim 3.5$). The addition of DPO is accompanied by an increase in the weight-average molecular mass of polyisoprene obtained in the presence of fraction I. Under a hydrodynamic action (method 2) on catalysts C-1 and C-2, polyisoprene with an increased weight-average molecular mass is formed. During the addition of piperylene to the catalyst (C-3), M_w of polyisoprene increases to values characteristic of C-2 formed via method 2. The formation of the catalytic system $TiCl_4$–$Al(iso\text{-}C_4H_9)_3$–DPO–piperylene via method 1 at −10°C results in substantial increases in the activity of the catalyst and the reactivity of the active centers of fraction II. The polymerization of isoprene on catalyst fraction III, regardless of the conditions of catalytic system formation (electron donor additives, exposure temperature, hydrodynamic actions), yields a low molecular mass polymer with a poly dispersity of 3.0–3.5.

TABLE 14.2 Molecular-Mass Characteristics of *cis*-1,4-polyisoprene. 1 and 2 are the Methods of Catalyst Preparation

C	τ min	Fraction I				Fraction II				Fraction III			
		$M_w \times 10^{-4}$		M_w/M_n		$M_w \times 10^{-4}$		M_w/M_n		$M_w \times 10^{-4}$		M_w/M_n	
		1	2	1	2	1	2	1	2	1	2	1	2
C-1	3	17.1		2.6		43.2	43.6	4.6	4.8	25.4	21.4	3.7	3.8
	20	56.3		4.2		46.5	57.5	4.0	5.9	23.7	22.6	3.4	3.7
	40	64.9		4.6		47.4	62.7	4.1	5.8	27.6	28.9	3.6	3.6
	90	72.2		5.1		54.3	61.4	4.3	6.2	26.8	22.4	3.7	3.6
C-2	3	24.3		3.3		39.5	54.8	4.3	4.9	21.1	22.3	3.8	3.7
	20	69.7		3.8		41.2	67.5	3.9	4.6	24.6	24.6	3.7	3.4
	40	78.4		3.7		42.6	72.8	4.4	4.3	26.3	22.4	3.3	3.5
	90	84.6		4.0		58.0	73.2	4.3	4.2	24.7	24.3	3.5	3.3
C-3	3					66.2	60.5	3.9	3.8	22.4	20.0	3.8	3.6
	20					75.4	71.1	3.7	3.5	28.1	21.3	3.7	3.7
	40					79.6	78.8	4.1	4.0	21.5	23.6	3.8	3.8
	90					81.6	79.6	3.9	3.8	23.2	26.8	3.6	3.3
C-4	3					81.2	72.7	3.6	3.7	25.4	20.3	3.8	3.6
	20					62.7	51.3	3.3	3.2	24.8	21.6	3.7	3.7
	40					57.6	44.3	3.1	3.2	23.2	26.8	3.6	3.6
	90					61.4	48.2	3.2	3.4	19.6	21.5	3.1	3.3

The solution of the inverse problem of the formation of the MWD in the case of cis-1,4-polyisoprene made it possible to obtain curves of the active site distribution over kinetic heterogeneity (Figs. 3–6). As a result of averaging of the positions of all maxima three types of polymerization site that produce isoprene macromolecules with different molecular masses were found:

Type A (lnM =10.7), type B (lnM = 11.6), and type C (lnM = 13.4).

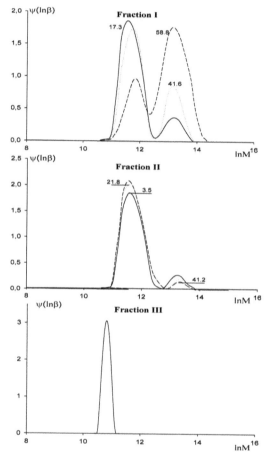

FIGURE 14.3 Active site distributions over kinetic heterogeneity during isoprene polymerization on fractions C-1. Method 1. Here and in Figs. 4–5 numbers next to the curves are conversions (%).

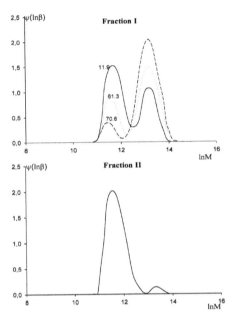

FIGURE 14.4 Active site distributions over kinetic heterogeneity during isoprene polymerization on fractions C-2. Method 1.

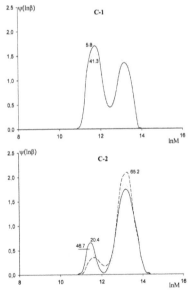

FIGURE 14.5 Active site distributions over kinetic heterogeneity during isoprene polymerization on particles of fraction II of C-1 and C-2. Method 2.

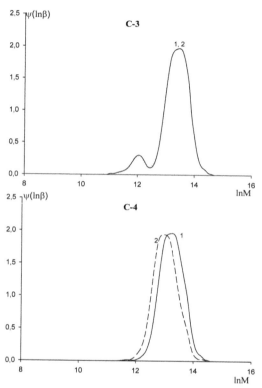

FIGURE 14.6 Active site distributions over kinetic heterogeneity during isoprene polymerization on particles of fraction II of C-3 and C-4. Method 1 – 1, method 2 – 2.

The polymerization of isoprene in the presence of fractions I and II of C-1 (method 1) occurs on site of types B and C (Fig. 14.3). The polymerization in the presence of fraction III proceeds on active site of type A only. The single site and low activity character of the catalyst composed of particles of fraction III is typical of all the studied catalysts and all the methods of their preparation. Thus, these curves are not shown in subsequent figures. Catalyst C-2 prepared via method 1 likewise features the presence of type B and type C site in isoprene polymerization on fractions I and II (Fig. 14.4). The hydrodynamic action (method 2) on C-1 accompanied by dispersion of particles of fraction I does not change the types of site of polymerization on particles of fraction II (Fig. 14.5). A similar trend is observed during the same action on a DPO-containing titanium catalyst in turbulent flows (Fig. 14.5).

In the presence of piperylene, the main distinction of function $\psi(\ln\beta)$ relative to the distributions considered above is a significantly decreased area of the peak due to the active site of type B (Fig. 14.6). The decrease of the catalyst exposition temperature to $-10°C$ (C-4, method 1) allows the complete "elimination" of active site of type B (Fig. 14.6). With allowance for the low content of fraction III, it may be concluded that, under these conditions, a single site catalyst is formed. In the case of the hydrodynamic action on C–4, the particles of fraction II contain active centers of type C with some shift of unimodal curve $\psi(\ln\beta)$ to smaller molecular masses (Fig. 14.6).

14.4 DISCUSSION

The particles of the titanium catalyst 0.03–0.14 µm in diameter, regardless of the conditions of catalytic system preparation, feature low activity in polyisoprene synthesis, and the resulting polymer has a low molecular mass and a narrow MWD. The molecular-mass characteristics of polyisoprene and the activity of the catalyst comprising particles 0.15–4.50 µm in diameter depend to a great extent on its formation conditions.

As shown in Refs. [10, 11], the region of coherent scattering for particles based on $TiCl_3$ spans 0.003–0.1 µm, a range that corresponds to the linear size of the minimum crystallites. Coarser catalyst particles are aggregates of these minimum crystallites. This circumstance makes it possible to suggest that the fraction of catalyst particles 0.03–0.14 µm in diameter that was isolated in this study is a mixture of primary crystallites of β-$TiCl_3$ that cannot be separated via sedimentation. The fractions of catalyst particles with larger diameters are formed by stable aggregates of 2–1100 primary crystallites. There is sense in the suggestion that the elementary crystallites are combined into larger structures via additional Al–Cl bonds between titanium atoms on the surface of a minimum of two elementary crystallites, that is, $(Ti)_1$–Cl–Al–Cl–$(Ti)_2$. Similar structures can be formed with the participation of AlR_2Cl and $AlRCl_2$, which are present in the liquid phase of the catalyst. Tri alkyl aluminum AlR_3 is incapable of this type of bonding. Thus, the structure of the most alkylated Ti atom (in the limit, a monometallic center of polymerization), which has the minimum reactivity, should be assigned to the active centers localized on particles 0.03–0.14µm in diameter [9]. On particles 0.15–4.50

μm in diameter in clusters of primary crystallites, high activity bimetallic centers with the minimum number of Ti–C bonds at a Ti atom are present. Thus, the experimental results obtained in this study show that the nature of the polymerization center resulting from successive parallel reactions between the pristine components of the catalytic system determines the size of the titanium catalyst particles and, consequently, their activity in isoprene polymerization.

14.5 CONCLUSION

We first examined the isoprene polymerization on the fractions of the titanium catalyst particles, which were isolated by sedimentation of the total mixture. The results obtained allow to consider large particle as clusters, which are composed of smaller particles. In the formation of these clusters are modified ligands available titanium atoms. In the process of polymerization or catalyst preparation the most severe effects are large particles (clusters). This result in the developing process later on substantially smaller as compared to initial size particles. These particles are fragments of clusters, which are located over the active centers of polymerization. Note that the stereo specificity is not dependent on the size of the catalyst particles.

Hypothesis about clusters agrees well with the main conclusions of this chapter:

I) Isolated the fraction of particles of titanium catalyst $TiCl_4$–Al(iso-$C_4H_9)_3$: I – 0.7–4.5 μm, II – 0.15–0.68 μm, III – 0.03–0.13 μm. With decreasing particle size decreases the rate of polymerization the molecular weight and width of the molecular weight distribution. Hydrodynamic impact leads to fragmentation of large particles of diameter greater than 0.5 μm.

II) Isoprene polymerization under action of titanium catalyst is occurs on three types active sites: type A – lnM = 10.7; type B – lnM = 11.6; type C – lnM = 13.4. Fractions I and II particles contain the active site of type B and C. The fraction III titanium catalyst is represented by only one type of active sites producing low molecular weight polymer (lnM = 10.7).

III) The use of hydrodynamic action turbulent flow, doping DPO and piperylene, lowering temperature of preparation of the catalyst allows to form single site catalyst with high reactivity type C (lnM = 13.4), which are located on the particles of a diameter of 0.15–0.18μm.

ACKNOWLEDGMENTS

This study was financially supported by the Council of the President of the Russian Federation for Young Scientists and Leading Scientific Schools Supporting Grants (project no. MD-4973.2014.8).

KEYWORDS

- Active sites
- Isoprene polymerization
- Particles size effect
- Single site catalysts
- Ziegler-natta catalyst

REFERENCES

1. Kissin, Yu. V. (2012). Journal of Catalysis 292, 188–200.
2. Hlatky, G. G. (2000). Chemical Reviews 100, 1347–1376.
3. Kamrul Hasan, A. T. M., Fang, Y., Liu, B., & Terano, M. (2010). Polymer 51, 3627–3635
4. Schmeal, W. R., Street, J. R. (1972). Journal of Polymer Science: Polymer Physics Edition 10, 2173–2183.
5. Ruff, M., & Paulik, C. (2013). Macromolecular Reaction Engineering 7, 71–83.
6. Taniike, T., Thang, V. Q., Binh, N. T., Hiraoka, Y., Uozumi, T., & Terano, M. (2011). Macromolecular Chemistry and Physics 212, 723–729.
7. Morozov, Yu. V., Nasyrov, I. Sh., Zakharov, V. P., Mingaleev, V. Z., Monakov, & Yu. B. (2011). Russian Journal of Applied Chemistry 84, 1434–1437.
8. Zakharov, V. P., Berlin, A., Monakov, A., Yu. B., & Deberdeev, R. Ya. (2008). Physicochemical Fundamentals of Rapid Liquid Phase Processes, Moscow: Nauka, 348 p.
9. Monakov, Y. B., Sigaeva, N. N., & Urazbaev, V. N. (2005). "Active Sites of Polymerization" Multiplicity: Stereospecific and Kinetic Heterogeneity, Leiden: Brill Academic, 397 p.

10. Grechanovskii, V. A., Andrianov, L. G., Agibalova, L. V., Estrin, A. S., & Poddubnyi, I. Ya. (1980). Vysokomol. Soedin., Ser. A 22, 2112–2120.
11. Guidetti, G., Zannetti, R., Ajò, D., Marigo, A., & Vidali, M. (1980). European Polymer Journal 16, 1007–1015.

CHAPTER 15

MODIFICATION OF RECEPTOR STATUS IN GROUPS OF PROLIFERATIVE ACTIVITY OF BREAST CARCINOMAS

A. A. BRILLIANT, S. V. SAZONOV, and Y. M. ZASADKEVICH

CONTENTS

ABSTRACT

The aim of the research was to evaluate modification of receptor profile of infiltrative breast carcinomas in accordance with dynamics of proliferative processes. The apportionment of the invasive carcinomas according to their proliferative activity was analyzed during the research. Dynamics of change of receptor status dependent on proliferative activity of the tumor was detected with the differentiated approach.

15.1 INTRODUCTION

It is known, that some prognostic and predictive factors should be considered to solve the problem of the therapy of breast carcinomas. A prognostic importance of the cell proliferation index (Ki67) is significant for those tumors, in which it is difficult to predict clinical course only with histological characteristics [1]. The cell proliferation index is an independent predictor of the general survival as well as the disease relapse in breast carcinoma patients [2]. Additionally, univariate and multivariate analyzes show that the cell proliferation index correlates with unfavorable clinical outcome [1, 3, 4].

During the breast carcinoma tissue examination, determination of steroid hormones receptors such as Estrogen (ER) and Progesterone (PR) is almost always used. It plays a key role in the correct assignment of hormonal therapy [5]. It is known that in three groups (ER+ PR+), (ER- PR+) and (ER+ PR-) resistance to hormone therapy (tamoxifen) is determined more often in the third group. It can be explained by the fact that in patients with "ER+ PR-" status the level of Her-2/neu and EGRF is higher [6].

Her-2/neu is a protooncogene, which encodes human epidermal growth factor receptor 2 (c-erb-2), related to tyrosine kinase group. Hyper expression of this oncogene is observed in 25–30% of breast cancer cases and associates with a poor prognosis in both with presence of metastases or without them [7].

The above listed tumor biomarkers such as Ki67, ER, PR, Her-2/neu are recommended for widespread clinical use nowadays [8]. The combination of them gives necessary information about receptor status of a tumor. However, dynamics of processes in a tumor remains still unclear. Study of dependence of receptor status from proliferation complicates with hetero-geneity of the explored tumors.

15.2   MATERIALS AND METHODS

Selected cases were analyzed by histological, immunohistochemical, mor-phometric and statistical methods. 406 cases of breast carcinomas were studied. Material for research was supplied by Sverdlovsk regional cancer center and Municipal Mammological Centre of Clinical Hospital № 40, Ekaterinburg, Russia. For immunohistochemical method glasses covered by POLYSINE SLIDES (Thermo scientific, Germany) were used. Her-2/neu expression at tumor cells was detected by polyclonal antibodies Poly-clonal Rabbit Anti-Human c-erb-2 Oncoprotein (DAKO, Denmark), Es-trogen and progesterone receptors expression was detected by monoclonal antibodies Monoclonal Mouse Anti-Human Estrogen Receptor, Monoclo-nal Mouse Anti-Human Progesterone Receptor (DAKO, Denmark). Level of proliferative activity was studied by evaluation of cell proliferation biomarker (Ki67) expression. For detection of the cell proliferation index antibodies Mouse Anti-Human KI-67 Antigen (DAKO, Denmark) were used. Proliferative activity of the investigated tumor can be evaluated by percentage ratio of stained nuclei of breast carcinoma cells to unstained. Immunohistochemical tests were made in autostainer "DAKO" (Denmark) with use of Dako Wash Buffer, visualization system Dako EnVision+Dual Link System-HRP, chromogen Dako Liquid DAB+ Substrate Chromogen System (DAKO, Denmark). Test evaluation was made with the robitic mi-croscope "Zeiss Ymager M" (Germany). Membrane expression of Her-2/neu in tumor cells was evaluated on the scale from 0 to 3+ [9]. Level of estrogen and progesterone receptors expression was detected on the scale from 0 to 8+ [10].

15.3   RESULTS AND DISCUSSION

After processing of 406 cases three groups of patients accordingly to Ki–67 protein expression were formed. The first group contained 248 cases (61% of the explored cases) in which Ki67 expression was lower than 10% inclusively. The second group contained 82 cases (20% of explored cases), in which Ki67 expression was higher than 10% but lower than 30%

inclusively. The third group contained 76 cases (19% of explored cases), in which Ki67 receptors were detected in more than 30% of tumor cells (Fig. 15.1). Division of cases of patients with breast carcinoma according to percentage of tumor cells expressed Ki67 is shown at the (Fig. 15.2). Thus, a majority of breast carcinomas has a low level of proliferation that is less than 10% of tumor cells express Ki67 protein receptors.

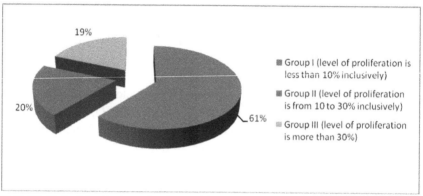

FIGURE 15.1 Distribution of patients with breast carcinoma according to a percentage ratio of cell expressing Ki–67.

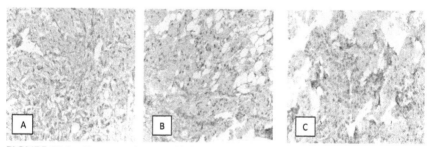

FIGURE 15.2 Breast carcinoma with level of proliferation A-10% of tumor cells, B-30% of tumor cells, C-50% of tumor cells. X100. Staining: IHC reaction HRP/DAB, additional staining with Mayer's haematoxylin.

Additional research, allowing evaluating features of receptor profile of a tumor according to its proliferative activity, was conducted (Table 15.1). After comparative analysis of 406 cases it was found that group I (degree of proliferation is less than 10%) does not have any significant difference in receptor status in comparison with group II (with degree of proliferation

10–30%). In turn, group II and group III (degree of proliferation is more than 30%) have some differences in their receptor profile. With increasing of proliferative activity of carcinoma cells expression of steroid hormones (Estrogen receptor, Progesterone receptor) is decreasing and Her-2/neu expression is increasing (Table 15.1).

TABLE 15.1   Receptor Status of Breast Carcinoma in Groups of Ki-67 Expression

Group of Ki-67 expression	Average definition of receptor status	
Group I – cases with low level of proliferation (less than 10% inclusively) n=248 (61%)	Estrogen receptor	2.5±0.2; p(1-2) >0.05
	Progesterone receptor	2.2±0.2; p(1-2) >0.05
	HER-2/neu	1.9±0.8; p(1-2) >0.05
Group II – cases with medium level of proliferation (more than 10%, less than 30% inclusively) n=82 (20%)	Estrogen receptor	2.3±0.2; p(2-3)<0.05
	Progesterone receptor	2.0±0.2; p(2-3)<0.05
	HER-2/neu	2.0±0.9; p(2-3) <0.05
Group III – cases with low level of proliferation (more than 30%) n=76 (19%)	Estrogen receptor	1.5±0.3; p(3-1) <0.05
	Progesterone receptor	1.3±0.2; p(3-1) <0.05
	HER-2/neu	2.8±1.5; p(3-1) <0.05

Due to heterogeneity of analyzed carcinomas, included in the groups of proliferation, it was decided to conduct a study of a degree of proliferation in combination with presence or absence of expression of steroid receptors and Her-2/neu. It was detected that only 13 cases (3%) out of all 406 cases had a positive expression of Estrogen, Progesterone and Her-2/neu at the same time. After analyzing of the division of these cases for groups of proliferative activity, we found a significant increase of steroid hormones expression from the first to the second group and decrease to the third group of proliferation (Table 15.2). We did not find any significant difference in Her-2/neu expression in the groups of proliferation. Thus, in this group another dependence of change of level of receptor expression in changing of proliferative processes was found in comparison with the general result.

TABLE 15.2   Receptor Status of Carcinoma in Groups of Ki-67 Expression with Positive Steroid Hormones Receptors and Her-2/Neu Expression

Group of Ki-67 expression	Average definition of receptor status	
Group I – cases with low level of proliferation (less than 10% inclusively)	Estrogen receptor	3.6 ± 0.8; $p(1\text{-}2) < 0.05$
n=6 (46%)	Progesterone receptor	4.0 ± 1.3; $p(1\text{-}2) < 0.05$
	HER-2/neu	2.3 ± 0.4; $p(1\text{-}2) > 0.05$
Group II – cases with medium level of proliferation (more than 10%, less than 30% inclusively)	Estrogen receptor	5.3 ± 2.3; $p(2\text{-}3) < 0.05$
	Progesterone receptor	5.0 ± 1.4; $p(2\text{-}3) < 0.05$
n=4 (30%)	HER-2/neu	2.2 ± 0.1; $p(2\text{-}3) > 0.05$
Group III– cases with low level of proliferation (more than 30%)	Estrogen receptor	3.9 ± 1.3; $p(3\text{-}1) > 0.05$
n=3 (23%)	Progesterone receptor	4.1 ± 2.3; $p(3\text{-}1) > 0.05$
	HER-2/neu	2.1 ± 0.1; $p(3\text{-}1) > 0.05$

The next step of our research was to study a dependence of level of estrogen receptor expression from level of activity of proliferative processes in the group of carcinomas positive only to estrogen receptor and negative to progesterone receptor and Her-2/neu. 60 appropriate cases (15% from all carcinomas) were explored. The comparative analysis showed that the level of estrogen receptor expression in the first group of proliferative activity was significantly lower, than in the second and the third groups (Table 15.3). Therefore, when the level of proliferation of tumors increases, Estrogen receptor expression grows as well. When levels of Progesterone receptor and Her-2/neu expression in the groups of research were compared, significant difference was not found. It is worth noting that the number of cases with high level of proliferation and positive Estrogen receptor expression is less in three times than cases with low proliferative activity.

TABLE 15.3   Receptor Status of Carcinoma in Groups of Ki-67 Expression with Positive Expression of Estrogen Receptor

Group of Ki-67 expression	Average definition of receptor status	
Group I – cases with low level of proliferation (less than 10% inclusively)	Estrogen receptor	3.5±0.2; p(1-2) <0.05
	Progesterone receptor	0.01±0.01; p(1-2) >0.05
n=27 (45%)	HER-2/neu	0.3±0.1; p(1-2) >0.05
Group II – cases with medium level of proliferation (more than 10%, less than 30% inclusively)	Estrogen receptor	4.5±0.3; p(2-3) >0.05
	Progesterone receptor	0.01±0.01; p(2-3) >0.05
n=24 (40%)	HER-2/neu	0.3±0.1; p(2-3)>0.05
Group III – cases with low level of proliferation (more than 30%)	Estrogen receptor	4.8±0.4; p(3-1) <0.05
	Progesterone receptor	4.1±2.3; p(3-1) >0.05
n=9 (15%)	HER-2/neu	0.60±0.04; p(3-1) >0.05

During the research of dependence of level of Progesterone receptor expression from level of proliferative activity, positive only to Progesterone receptor and negative to Estrogen receptor and Her-2/neu, we found 32 cases (8% from all the breast carcinomas) appropriate under these criteria. We did not find any significant difference of expression of these receptors in the groups of proliferative activity; hence a dependence of progesterone receptor expression from proliferative activity of the tumor was not detected (Table 15.4). Her-2/neu expression in the groups of proliferative activity does not have any significant difference. Estrogen receptor expression does not change with the increase of the level of proliferation of the tumor. Number of cases of the third group (with high level of proliferation) is 3 and 6 times lower than number of cases in the second and the first groups of Ki67 expression respectively.

TABLE 15.4   Receptor Status of Carcinoma in Groups of Ki-67 Expression with Positive Expression of Progesterone Receptor

Group of Ki-67 expression	Average definition of receptor status	
Group I – cases with low level of proliferation (less than 10% inclusively)	Estrogen receptor	0.02±0.02; p(1-2) >0.05
	Progesterone receptor	4.3±0.3; p(1-2) >0,05
n=19 (60%)	HER-2/neu	0.01±0.00; p(1-2) >0.05

TABLE 15.4 *(Continued)*

Group of Ki-67 expression	Average definition of receptor status	
Group II – cases with medium level of proliferation (more than 10%, less than 30% inclusively)	Estrogen receptor	0.04±0.04; p(2-3) >0.05
	Progesterone receptor	4.3±0.4; p(2-3) >0.05
n=10 (30%)	HER-2/neu	0.09±0.06 p(2-3)>0.05
Group III – cases with low level of proliferation (more than 30%)	Estrogen receptor	0.0±0.0; p(3-1) >0.05
	Progesterone receptor	3.8±0.8; p(3-1) >0.05
n=3 (10%)	HER-2/neu	0.6±0.2; p(3-1) >0.05

From all 406 cases of infiltrative carcinoma 129 (32%) express Estrogen receptor, Progesterone receptor and negative to Her-2/neu at the same time. During the research of dependence of level of steroid hormones receptors expression on the level of proliferative activity of breast carcinoma in this group no significant difference in the change of proliferative activity was not found (Table 15.5). A small number of studied carcinomas related to the first group of proliferative activity. The more proliferative processes in a tumor are, the less number of cases with positive estrogen and progesterone receptors are. Number of cases decreases with the increase of proliferative activity from the first to the third group in 4 times.

TABLE 15.5 Receptor Status of Carcinoma in Groups of Ki-67 Expression with Positive Expression of Estrogen and Progesterone Receptors

Group of Ki-67 expression	Average definition of receptor status	
Group I – cases with low level of proliferation (less than 10% inclusively)	Estrogen receptor	4.3±0.1 p(1-2) >0.05
	Progesterone receptor	4.8±0.2; p(1-2) >0.05
	HER-2/neu	0.24±0.03; p(1-2) >0.05
n=62 (48%)		
Group II – cases with medium level of proliferation (more than 10%, less than 30% inclusively)	Estrogen receptor	4.8±0.2; p(2-3) >0.05
	Progesterone receptor	4.9±0.2; p(2-3) >0.05
	HER-2/neu	0.24±0.04; p(2-3)>0.05
n=50 (39%)		
Group III – cases with low level of proliferation (more than 30%)	Estrogen receptor	5.0±0.3; p(3-1) >0.05
	Progesterone receptor	5.3±0.3; p(3-1) >0.05
n=17 (13%)	HER-2/neu	0.2±0.1; p(3-1) >0,05

Forty-one cases (10% from all carcinomas) were included in the group of carcinomas, expressed Her-2/neu receptor (2+ and 3+ cases) and which were negative to Estrogen and Progesterone receptors. A half of all the cases was included in the second group of Ki67 expression and had a medium level of proliferation. We did not find any significant difference in Her-2/neu expression in the groups of proliferation thus dependence of Her-2/neu from proliferative activity of a tumor was not detected (Table 15.6).

TABLE 15.6 Receptor Status of Carcinoma in Groups of Ki-67 Expression with Positive Expression of Her-2/neu Receptor

Group of Ki-67 expression	Average definition of receptor status	
Group I – cases with low level of proliferation (less than 10% inclusively) n=9 (23%)	HER-2/neu	2.4 ± 0.1; $p(1\text{-}2)>0.05$
Group II – cases with medium level of proliferation (more than 10%, less than 30% inclusively) n=23 (54%)	HER-2/neu	2.5 ± 0.1; $p(2\text{-}3)>0.05$
Group III – cases with low level of proliferation (more than 30%) n=9 (23%)	HER-2/neu	2.3 ± 0.1; $p(3\text{-}1)>0.05$

We conducted an additional research allowing studying distribution of the cases with negative expression of Estrogen, Progesterone receptors and Her-2/neu to the groups of proliferation. In the result of studying of 64 cases (16%) it was found that 25% and 28%, respectively related to the groups with low and medium levels of proliferation. The majority of cases were attributed to the third group (with high level of proliferation – 44% from all the cases).

15.4 CONCLUSION

After analyzing the received data, it can be concluded that the general study, which does not consider heterogeneity of properties of the tumors included in the study, could be used for the evaluation of receptor profile of breast carcinomas. Thereby, distribution to the groups of proliferation of all the cases showed that steroid hormones receptors expression decreased and Her-2/neu expression increased with increasing of proliferative activity of breast carcinoma cells. Differentiated approach at the research of receptor status in the groups of tumor proliferation showed that number of tumors with high proliferative activity grows with the increase of Estrogen receptor expression. We did not find any significant difference in Progesterone receptor expression in change of level of proliferation in a tumor. Moreover, a majority of positive at Estrogen and Progesterone receptors expression cases related to the group with low level of proliferation. It was not also find any significant difference at Her-2/neu expression in the different groups of proliferative activity. The group with the medium level of proliferation included twice more cases of carcinomas with positive expression of Her-2/neu.

KEYWORDS

- **Breast carcinoma**
- **Immunohistochemistry**
- **Proliferation**
- **Receptor status**

REFERENCES

1. Scholzen, T. (2000). The Ki-67 protein: from the known and the unknown. J. Cell. Physio. (182), 311.
2. Penault-Llorca, F., Cayre, A., Bouchet Mishellany, F., et al. (2003). Induction chemotherapy for breast carcinoma: predictive markers and relation with outcome. Int. J. Oncol. (22), 25, 1319.
3. Gonzalez-Vela, M. C., Garijo, M. F., Fernandez, F., & Val-Bernal, J. F. (2001). MIB1 proliferation index in breast infiltrating carcinoma: comparison with other proliferative markers and association with new biological prognostic factors. Histol. Histopathol. (16), 399–406.
4. Jones, S., Clark, G., Koleszar, S, et al. (2001). Low proliferative rate of invasive node negative breast cancer Predicts for a favorable outcome: a prospective evaluation of 669 patients. Breast Cancer (1), 310–314.
5. Sazonov, S. V., & Leontiev, S. L. (2012). Creation of the system of revision of immunohistochemical research in breast cancer diagnostics. Vest. of Ural Academ. Science, 38(1), 18–23.

6. Garin, A. M., & Tver, M. (2005). Endocrine therapy and hormone dependent tumors, Triada. 240 p.

7. Joerger, M., Thürlimann, B., & Huober, J., (2011 Jan; 22). Small HER2-positive, node-negative breast cancer: who should receive systemic adjuvant treatment? Ann. Oncol. 1, 17–23.

8. Frank, G., Zavalishina, L., Andreeva, J., Matsionis, A., & Sazonov, S. (2012). HER2 testing in Russia: The results of the 10 years of experience. ASCO/CAP recommended score system Wirchows Archive, (461) (Suppl 1), 241–242.

9. Bilous, M., Dowsett, M., Isola, J., Lebeau, A., Moreno, A, et al. (2003). Current perspectives on HER2 testing: a review of national testing guidelines Mol. Pathol. (16), 82–173.

10. Elledge, R. M., Green, S., Pugh, R., Allred, D. C., Clark, G. M., Hill, J, et al. (2000). Estrogen receptor (ER) and Progesterone receptor (PgR), by ligand-binding assay compared with ER, PgR and pS2, by immunohistochemystry in predicting response to tamoxifen in metastatic breast cancer: a South-west Oncology Group Study. Int. J. Cancer. (89), 7–11.

CHAPTER 16

SORBTION PROPERTIES OF BIODEGRADABLE POLYMER MATERIALS BASED ON LOW-DENSITY POLYETHYLENE, MODIFIED CHITOSANS

MARINA BAZUNOVAA, IVAN KRUPENYAA, ELENA KULISHA, and GENNADY ZAIKOV

CONTENTS

ABSTRACT

Compositions obtained ultrafine powders based on low-density polyethylene, a modified natural polymer chitosan under the combined influence of high pressures and shear deformation. Studied their sorption properties and susceptibility to biodegradation. The resulting samples of polymer films based on low-density polyethylene modified chitosan having acceptable strength characteristics, good absorbent capacity and biodegradability can be used for manufacturing biodegradable packaging materials.

16.1   INTRODUCTION

For materials for the manufacture of food packaging, disposable products, it is advisable the use of biodegradable polymers, which retain only the performance during the period of consumption, and then undergo a physicochemical and biological transformations under the influence of environmental factors, and easily incorporated into natural metabolic processes biosystems.

The problem of biodegradability of well-known tonnage industrial polymers is quite urgent for modern studies. It is promising enough to use synthetic and natural polymer mixtures, which can play the roles of both filler and modifier for creating biodegradable environmentally safe polymer materials. The macromolecule fragmentation of the synthetic polymer is to be provided for due to its own bio-destruction.

The synthetic polymers have been modified by the natural one under the combined effect of high pressure and shear deformation. The usage of this method for obtaining polymer composites is sure to solve several problems at once. Firstly, the ultra dispersed powders with a high homogeneity degree of the components can be obtained under combined high pressure and shear deformation thus resulting in easing the technological process of production [1]. Secondly, the elastic deformation effects on the polymer material may lead to the chemical modification of the synthetic polymer macromolecules by the natural polymer blocks via recombination of the formed radicals. Thus, it can provide for the polymer product biodegradation.

As components in the preparation of biodegradable polymeric composite synthetic polymer used a large-capacity, low density polyethylene (LDPE) and naturally occurring polysaccharide chitosan (CTZ).

The parameters characterizing the tendency of the compositions to biodegradation selected by their ability to absorb water (degree of swelling) and the mass loss by maintaining samples in soil. Water absorption is one of the indirect indicators of the propensity of the material to biodegradation, as the swollen material accelerated diffusion processes synthesized by microorganism's enzymes catalyzing the process of biological degradation [2].

Therefore, is appropriate to study the sorption properties and ready biodegradability compositions based on ultrafine powders LDPE modified natural polymers HTZ under the combined effects of high-pressure and shear deformation.

16.2   EXPERIMENTAL PART

LDPE 10803-020 (90000 molecular weight, 53% crystallinity degree, and 0.917 g/sm^3 density) and chitosan samples of Bioprogress Ltd. (Russia) obtained by alkalinedeacetylation of crab chitin (deacetylation degree ~84%), and M_{sd}= 115,000 were used as components for producing biodegradable polymer films.

The initial highly dispersed powders with different mass ratio of components have been obtained by high temperature shearing (HTS) under simultaneous impact of high pressure and shear deformation in an extrusion type apparatus with a screw diameter of 32 mm [3, 4]. Temperatures in kneading, compression and dispersion chambers amounted to 150°C, 150°C and 70°C, respectively.

The size of particles in powders of LDPE, CTZ and LDPE/CTZ with various mass ratio of the components were determined by "Shimadzu Salid–7101" particle size analyzer. The film formation was carried out by rotomolding [5] at 135 and 150°C. The film sample thickness amounted to 100 μm and 800 μm.

As a measure of the degree of modification (P, HTZ grams to 1 gram polyethylene) adopted mass HTZ who "grafted" onto polyethylene. Here it is assumed that HTZ chemically bound to polyethylene, insoluble in 1% acetic acid solution, unlike the unbound HTZ.

The absorption coefficient of the condensed vapors of volatile liquid (water, n-heptane) K' in static conditions is determined by complete saturation of the sorbent by the adsorbent vapors under standard conditions at 20°C [6] and was calculated by the formula: $K' = \frac{m_{absorbed\ water}}{m_{sample}} \times 100\%$, where $m_{absorbed\ water}$ is weight of the saturated condensed vapors of volatile liquid, g; m_{sample} is weight of dry sample, g.

Film samples were long kept in the aqueous and enzyme media to determine the water absorption coefficient while the absorbed water weight was calculated. The water absorption coefficient of film samples of LDPE/CTZ with different weight ratio was determined by the formula: $K = \frac{m_{absorbed\ water}}{m_{sample}} \times 100\%$, where $m_{absorbed\ water}$ is water weight absorbed by the sample whereas m_{sample} is the sample weight. Sodium aside was added to the enzyme solution to prevent microbial contamination. Each three days both the water medium and the enzyme solution were changed. The "Liraza" agent of 0.1 g/L concentration was used as an enzyme (Immunopreparation SUE, Ufa. Russia). In experiments for determining the absorption of the condensed vapors of volatile liquid and water absorption coefficients at a confidence level of 0.95 and 5 repeated experiments, the error does not exceed 7%.

In assessing the activity of composites to bind the protein as a biological marker albumin solution used in the model [7] obtained by precipitating casein from nonpasteurized milk, followed by separating the casein by centrifugation. The concentration of albumin to sorption and thereafter determined spectrophotometrically according to the formula: Protein content = $1.45 \times D_{280} - 0.74 \times D_{260}$ (mg/mL) wherein D_{280} – absorbance of the solution at 280 nm; D_{260} – absorbance of the solution at 260 nm.

The obtained film samples were kept in soil according to the method [8] to estimate the ability to biodegradation. The soil humidity was supported on 50–60% level. The control of the soil humidity was carried out by the hygrometer ETR–310. Acidity of the soil used was close to the neutral with pH = 5.6–6.2 (pH-meter control of 3in1 pH). At a confidence level 0.95 and 5 repeated experiments the experiment error in determining the tensile strength and elongation does not exceed 5%.

Mechanical film properties (tensile strength (σ) and elongation (ε)) were estimated by the tensile testing machine ZWIC Z 005 at 50-mm/min tensile speed.

16.3   RESULTS AND DISCUSSION

It can be assumed that the process of elastic-strain effects on the mixture of LDPE and CTZ can lead to chemical modification of macromolecules synthetic polymer blocks natural polymer due to the recombination of macroradicals formed. In this regard, the evaluated degree of modification of polyethylene by treating chitosan superfine powders LDPE / CTZ obtained by HTS, an excess of 1%–solution of acetic acid. When it is allowed that chitosan chemically linked polyethylene, is insoluble in 1% acetic acid solution, unlike the unbound chitosan. Data on the extent of modification of polyethylene chitosan (P) and on the proportion of (D) CTZ, which entered the process in the modification with polyethylene on its total weight, are shown in Table 16.1.

From Table 16.1, it follows that the combination of the initial components under HTS leads to a fairly high degree of modification of chitosan macromolecules polyethylene fragments.

The speed of the hydrolytic destruction of the polymer materials is closely connected with their ability to water absorption. Values of their absorption capacity according to water and heptane vapors were determined for a number of powder mixture samples of LDPE/CTZ (Table 16.2). It was established that the absorption coefficient of the condensed water vapors is directly proportional to the chitosan content.

As the initial powders, the films with high chitosan content under rotomolding absorb water well (Table 16.3). At the same time thinner films absorb more water for a shorter period of time.

TABLE 16.1   The Degree of Modification of Polyethylene Chitosan (P) and Share (D) HTZ, Entered Into the Process of Modifying Polyethylene During HTS

№	LDPE/CTZ powder, mass. %	Particle size, μm	P, g HTZ/1 g LDPE	D, %
1	0	5.5–8.0; 10.0–80.0	–	–
2	20	6.5–63.0	0.19	25.7
3	40	6.5–50.0	0.32	38.3
4	50	4.3–63.0	0.45	44.6
5	60	6.5–63.0	0.73	48.4

TABLE 16.2 The Absorption Coefficient of the Condensed Water Vapors of Volatile Liquid (water and n-heptane) K' of LDPE/CTZ powders at 20 °C

№	LDPE/CTZ powder, mass. %	K' by water vapors, %	K' by n-heptane, %
1	0	1.10±0.08	17±1.0
2	20	12.3±0.8	11.0±0.8
3	40	20±1.0	5.0±0.4
4	50	25±2.0	4.0±0.3
5	60	35±2.0	4.0±0.3

TABLE 16.3 Values of Equilibrium Water Absorption Coefficients K (%) of LDPE/CTZ films at 20°C

№	LDPE/CTZ powder, mass. %	K, %			
		Medium - water		Medium – liraza enzyme (0,1 g/L)	
		Film thickness 100 μm	Film thickness 800 μm	Film thickness 100 μm	Film thickness 800 μm
1	20	5.0±0.4	2.0±0.2	5.0±0.4	4.0±0.3
2	40	10.0±0.7	4.0±0.3	13.0±0.9	7.0±0.5
3	50	38±3	14±1.0	40±3.0	45±3.0
4	60	–	31±2.0	–	95.8±0.7

In case the film samples were placed into the enzyme solution, water absorption changes slightly. Firstly, the equilibrium values of the absorption coefficient of films in the enzymatic medium are higher than in water (Table 16.3). It is in the enzymatic medium usage that a longer film exposure (for more than 30–40 days) was accompanied by weight losses of the film samples. Moreover, after 40 days of testing, the film with 50%mass of chitosan and 100um thickness lost its integrity. Films of 800 μm thick and chitosan content of 50 and 60% lost their integrity after 2 months of the enzyme agent solution contact (Fig. 16.1). These facts are quite logical as "Liraza" is subjected to a β-glycoside bond break in chitosan. Thus, the destruction of film integrity is caused by the biodestruction process. Higher values of the water absorption coefficient may be explained by enzyme destruction of chitosan chains as well due to some loosening in the film material structure (Table 16.3).

In assessing the activity of composites to bind the protein as a biological marker albumin solution used in the model. Found that the maximum of activity to bind a protein after exposure of samples in the model solution for 24 h have LDPE film / HTZ composition 40/60 wt. % Lowering serum albumin by about 17%. Consequently, the composites obtained showing sorption activity in relation to the nature of biological markers.

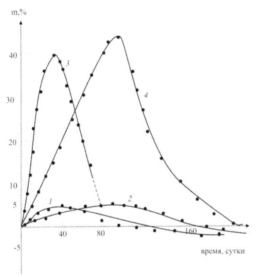

FIGURE 16.1   Curves of relative mass change of the film samples LDPE/ CTZ, immersed in a solution of the enzyme preparation "Liraz" concentration of 0.1 g/L (20 °C): (1) CTZ content of 20 wt.%, the film thickness of 100 μm, (2) the contents CTZ 20 wt. %, the film thickness of 800 μm, (3) the contents CTZ 50 wt.%, the film thickness of 100 μm, and (4) the contents CTZ 50 wt.%, the film thickness of 800 μm.

Tests on holding the samples in soil indicate on bio-destruction of the obtained film samples either. It is found that the film weight is reduced by 7–8% during the first six months. Here the biggest weight losses are observed in samples with 50–60 mass % of chitosan.

Chitosan introduction into the polyethylene matrix is accompanied by changes in the physical and mechanical properties of the film materials. The polysaccharide introduction into the LDPE compounds results in slight decrease in the tensile strength of films. Wherein the number of the chitosan introduced does not affect the composition strength. However, low-density polyethylene/chitosan films obtain much less elongation values as compared with low-density polyethylene films under the same conditions. Thus, films, which were obtained on the basis of ultra dispersed LDPE powders modified by chitosan possess less plasticity while, retain in got her satisfactory strength properties.

Thus, are obtained composition based on the ultrafine powders LDPE modified HTZ under the combined influence of high-pressures and shear deformation. Samples of polymer films based on LDPE modified HTZ having acceptable strength characteristics, good absorbent capacity, and bio-degradability can be used for manufacturing biodegradable packaging materials.

KEYWORDS

- **Biodegradable Polymer Films**
- **Chitosan**
- **Low Density Polyethylene**
- **The Water-Absorbent Capability**

REFERENCES

1. Bazunova, M. V., Babaev, M. S., Bildanova, R. F., Yu, A., Protchukhan, S., Kolesov V., & Akhmetkhanov, R. M. (2011). Powder-polymer technologies in sorption-active composite materials *Vestn. Bashkirs. Univer,* 16(3), 684–688.
2. Fomin, V. A., & Guzeev, V. V. (2001). Biodegradable polymers, condition and prospects. *Plasticheskie Massi.* 2, 42–46.
3. Enikolopyan, N. S., Fridman, M. L. Yu, A. Karmilov. (1987). Elastic-deformation grinding of thermo-plastic polymers. *Reports AS USSR,* 296(1), 134–138.
4. Akhmetkhanov, R. M., Minsker, K. S., Zaikov, G. E. (2006). On the mechanism of fine dispersion of polymer products at elastic deformation effects. *Plasticheskie Massi,* 8, 6–9.
5. Sheryshev, M. A. (1989). *Formation of polymer sheets and films.* (Ed). Braginsky, V. A., Leningrad: Chemistry Publishing, 120 p.
6. Keltsev, N. V. (1984). *Fundamentals of adsorption technology.* Moscow: Chemistry, 595 p.

7. Asher, W. J., Davis, T. A., & Klein, E. (1989). *Sorbents and their clinical application.* (Ed.) Giordano, C. Kiev: Naukova Dumka, 398 p.
8. Ermolovitch, O. A., Makarevitch, A. V., Goncharova, E. P., & Vlasova, F. M. (2005). Estimation methods of biodegradation of polymer materials. *Biotechnology*, 4, 47–54.

CHAPTER 17

QUANTITATIVE ASSESSMENT OF FUNGICIDAL AND BACTERICIDAL ACTIVITY OF NANOSTRUCTURAL SILVER PARTICLES

I. G. KALININA, K. Z. GUMARGALIEVA, V. P. GERASIMENYA, S. V. ZAKHAROV, M. A. KLYIKOV, and S. A. SEMENOV

CONTENTS

ABSTRACT

It is shown that injection of silver nanoparticles (NPs) into large-tonnage polymers, such as polystyrene (PS) and styrene copolymer with acrylonitrile (SAN), imparts fungicide properties to them. For the purpose of further forecasting, the process of microscopic fungi and bacteria growth in the presence of different silver NP concentrations has been described quantitatively. It is shown that addition of silver NPs significantly suppresses growth of Aspergillus niger and Penicillium chrysogenum microscopic fungi both at the initial, and stationary stages of their growth. Inhibition of bacterial growth manifests itself in increased induction period (the lag-phase).

17.1   INTRODUCTION

It is common knowledge that silver ions are highly toxic for both microorganisms and bacteria, for instance, *E. coli* [1, 2]. It has been shown [3] that silver NPs inhibit microbial growth, suppressing it by the free-radical mechanism that gives an opportunity to use them in medicine and antimicrobial control systems. Meckling et al. [4] have shown that silver NP compounds with hyper-branched amphiphilic macromolecules are highly efficient antimicrobial agents, too.

Silver NPs (1–2 nm) hybrids with modified highly branched amphiphilic polyethylene imines effectively adhere to polar surfaces imparting them the antimicrobial property. It is well known that colloidal silver manifests an antimicrobial property, but its particles adhere poorly to the surface. The authors have synthesized hybrids of silver particles and modified highly branched amphiphilic polyethylene imine that gives an antimicrobial coating. Obtaining of silver NPs and the effect of medium components on their formation in a composite solvent of methyl cello solve-butyl acetate-toluene and in solution of methyl metacrylate copolymer with metacrylic acid has been considered in Ref. [5]. Using the atomic-force microscopy method, it is shown that the particles sized within 50–500 nm are formed in the silver trifluoroacetate composite solution system. The authors have shown that butyl acetate and toluene additions increase stability of silver ion-methyl cello solve complexes. Copolymer molecules in the solution prevent enlargement of silver NPs thus decelerating their precipitation [5]. Alongside with that, despite numerous studies performed in this field, the mechanism of silver NPs action is not unanimous. The most researchers associate microorganism growth suppression with free radicals formation on the silver NP surface, which attack the microorganism cell membrane and destroy it completely [4]. In the opinion of another group of authors, the inhibiting action mechanism of nano-sized silver particles may conclude in the electrostatic effect of attraction between negatively charged cell membrane of the microorganism and positively charged silver NP [6–8]. On the contrary, Sondi and Salopek [8] point out that antimicrobial silver NP activity for gram-negative bacteria depends on their concentration and is associated with formation of "plagues" on the bacterial cell membrane that change permeability of the cell membrane and cause the cell death due to accelerated effluence of lipopolysaccharide and protein molecules from the membrane [9, 10]. Antimicrobial activity of silver NPs is also associated [3, 11] with the impact of radicals on cellular membranes shown by the ESR method. Electron microscopic analysis has clearly shown that silver NPs accumulate in the membrane, because some of them successfully permeate into the cell.

Bleeding of intercellular substances and coagulation of nano-sized particles on the bacterial surface can be seen on a gating microscope. Works by Klabunde [12] have demonstrated that NPs of active metal oxides show high antibacterial activity and thus it is of interest to study the use of other inorganic NPs as antibacterial materials. Little is known about the biocide effect of precious metal particles. The mechanism of silver ions inhibitor effect is partly known. It is suggested that DNA loses its reproductive ability and cellular proteins become inactive, when treated by silver ions [13]. It is also shown that silver ion bonds with functional groups of proteins cause degradation of proteins [14]. The features of surfactant catalytic action in hydrocarbons and lipids oxidation have been considered [15, 16]. It is shown that hydro peroxides, the primary amphiphilic products of oxidation of lipids, form mixed micelles with surfactants, in which rapid decay of peroxides happen; other polar components (metal-containing compounds, inhibitors, etc.) are accumulated that significantly affects oxidation rate and mechanism. Comparison of surfactant action of different origin has shown that cationic surfactants speed up hydro peroxide decay with free radical formation [18], that is, are catalysts of hydro peroxide radical decay. Anionic and nonionic surfactants have no such influence. Hydro peroxides decay into radicals provides degenerated branching of chains and general autocatalytic development of the oxidation process. The mechanism of cationic surfactants catalytic effect on oxidation processes involves acceleration of degenerated branching of chains in the course of hydro peroxides decay. Generated peroxy radicals pass to the volume and may initiate chain oxidation. Hydro peroxides are also formed in living organisms during biochemical processes. In the presence of cationic surfactants, catalytic degradation of lipoperoxide cellular membranes into radicals and further radical reactions with polyene compounds, lipids, proteins and other components in the cell, which cause their irreversible degradation, represent the possible mechanism of bactericide action of cationic surfactants.

It has been shown [19] that silver is sorbed well by a broad range of microorganisms: algae, fungi and bacteria. However, the most works on silver interaction with cells are devoted to its action in the ionic form [20]. Of interest are works on biological action of silver NPs on yeast cells. The interaction between ions and stable nanosized silver clusters synthesized in inverted micelles by the radiation-chemical method have been studied in a wide range of concentrations on Candida utilize and Saccharomyces Cerevisiae yeast cells in aqueous and aqueous-organic solutions [19]. It has been found that the biocide effect of Ag clusters exceeds the effect of silver ions. It is shown that Ag ions have no effect on yeast cell growth, whereas NPs suppress fermentation. It is also shown [21] that the concentration-dependent toxic effect of ions in relation to bacteria and yeasts is associated with the binding of Ag ions to proteins and lipids of cellular membranes and subsequent change of the transmembrane potential up to membrane breakdown of the cell death. The mechanism of silver NP effect of living cells remains unclear. The inhibiting effect of silver NPs on development of a number of microorganisms is shown. Silver NP organization, both outside the cell and in the preplasmatic space (in bacteria) or on the cell wall surface (in yeasts), is another result of silver ions interaction with microorganisms (at silver ions concentration above 45×10^{-6} M). Concurrently with investigations of the mechanism of silver or other metals NP formation, the branch of nanoscience and nanotechnology that uses achievements of physics, chemistry, engineering and technics at the nanoscale levels is ever expanding. In the recent decade, the number of researches in the field of nanotechnologies and nanocomposites (polymer-nanoparticles composites) grows exponentially, and special attention is paid to the "structure-properties" interconnection and its application.

Degradation and service life of polymers, as well as the NP role in biodegradation of polymers are considered in the presence of NPs (nanocompo-
nents) under various environmental conditions [23, 24].

For research and applied works, there are three main directions: biodegradable polymers based on polyesters of hydroxy carboxylic acids, composite materials based on natural polymers, modification of existing industrial polymers and attribution of novel properties to them. Large-tonnage
polymers: PE, PP, PVC and PS without modifications and articles from
them can be stored for decades. Sometimes articles require antibacterial
properties, but after the end of service life they have to be disposed of.
Wide application of biodegradable polymers is embarrassed because of
their high cost as compared with traditional polymers.

When obtaining new materials, it is often desirable, if not particularly
enhancement, but at least preservation of their physicochemical and bacterial properties that may be reached by using a nano-additive to the polymer
at the processing. The goal of this work was to inject silver NPs into large-
tonnage polymers, polystyrene (PS) and styrene copolymer with acrylonitrile (SAN), and to determine antimicrobial and fungicide properties of
polymers comprising silver NPs, and quantitatively specify microbiological overgrowth of the polymers for forecasting purposes.

Production of new composite materials provokes continued scientific
interest in various types of their degradation and strength. As we have
shown in our studies of bio-stability of polymers and metals [25–27], materials are biologically damaged at their contact with the living organisms
that causes changes in their performance. In general, the following processes proceed during bio-damaging: adsorption of microorganisms on the
material surface; growth of microorganisms; material degradation as a result of either specific action (living organisms consume a polymeric material as a nutrition) or the action of metabolic products. Growth and development of microscopic fungi and bacteria on solid surfaces is commonly
evaluated by the six-grade scale using the GOST approved methods or by
colony diameter increase for a definite kind or set of microscopic fungi.
This is because of experimental difficulties in detection on the biomass
amounting several micrograms per cm^2 at the initial growth stages [28]. To
protect materials against biofouling low-molecular chemical substances,
the so-called biocides, are used. The list of substances possessing biocide
properties expands continuously. At present, methods of evaluation of fungicide activity of chemical substances based on measurements of fungus

colony growth rate on agarized media in the presence of these substances [28] are widely applied. This method is semiquantitative and subjective and does not allow determination of the influence of biocides on various stages of microorganisms development. Since biofouling of materials develops with time, kinetic methods of investigation are the maximum extent suited to evaluation of biocide efficiency [29, 30]. The goal of this work is in quantitative assessment of growth of GOST approved species of microscopic fungi and bacteria in the presence of multifunctional modifying additive (MMA) of "Akvivon-TM," TU 2499–024–87552538–12 with various silver NP concentrations and evaluation of its influence on physicochemical properties of polystyrene and styrene copolymer with acrylonitrile, as well as on the possibility of adhesion and growth of bacteria and fungi on PS and SAN surfaces.

17.2   MATERIALS AND METHODS

Fungicide properties of polymeric samples with silver NPs applied on polymeric granules from 2% aqueous colloid solution of silver NPs in organic dispersion of "AKVIVON" brand, TU 2499-022-87552538-10, or without them were studied on polymeric disks from polystyrene and styrene copolymer with acrylonitrile. The disks were press molded at 160°C and then cooled down to 60°C during 40-60 min. The samples are 100 μm thick and 50 mm in diameter. Temperature characteristics of the samples, glass transition temperature (T_g), in particular, were determined on DSC Q100 calorimeter by company TA Instruments. Table 17.1 shows results of direct and repeated melting.

TABLE 17.1   Glass Transition Temperatures for Direct and Repeated Melting For Polystyrene, Styrene Copolymer with Acrylonitrile in the Presence and in the Absence of Silver Nanoparticles

Sample name	T_g, °C	T_g, °C, Repeated melting
Styrene copolymer with acrylonitrile, control	110.4	110.6
Styrene copolymer with acrylonitrile with silver NPs, version 1	111.0	110.7
Polystyrene, control	97.9	97.4
Polystyrene with silver NPs, version 2	111.0	110.7

The presence of silver complexes in polymers had no effect on IR-Fourier spectra of all samples, that is, on temperature and spectral characteristics of samples.

According to GOST 9.049-91, purified polymeric samples were placed into Petri dishes on a solid nutrient medium (the Chapek-Dox medium with agar), contaminated by a suspension of fungi spores in the Chapek-Dox medium and exposed to conditions optimal for development of fungi, and then fungicide properties was evaluated by intensity of fungus growth on the samples and the nutrient medium.

Test conditions: duration 14 days; constant temperature of $+29\pm2°C$; relative air humidity over 90%.

Mold fungi species: *Aspergillus niger van Tieghem, Aspergillus terreus Thom, Aspergillus oryzae (Ahlburg), Penicillium funiculosum Thom, Penicillium chrysogenum Thom, Penicillium cyclopium Westling, Paecilomuces varioti Bainier, Chaetomium globosum Kunze, Trichoderma viride Pers. Ex Fr.*

Indices of fungicide properties are growth intensity of mold fungi in samples in points by the six-grade scale, GOST 9.048-89 [31]; the presence of the inhibitor zone (the growth absence zone) in the nutrient medium around the sample. According to GOST 9.049–91 [32], a strong fungicide effect is characterized by the absence fungi growth on the sample (point 0). The absence of growth on the sample and the presence of the inhibitor zone on the nutrient medium around the sample mean a strong manifestation of the fungicide effect of silver NPs as a result of diffusion into the nutrient medium. Growth of fungi on the sample that corresponds to point 1, indicates low fungal resistance of the material, whereas points 2–5 growth indicates the absence of the fungicide effect.

17.3   RESULTS AND DISCUSSION

The results of the investigation indicate that PS and SAN samples with silver NPs manifest high fungicide properties. At the same time, control samples have no fungicide properties [32].

Electron microscopic analysis of polymer samples after rinsing microbial colonies from their surfaces indicates that adhesion is irreversible, in the presence of silver NPs numerous fragments of microorganisms being observed.

17.3.1 TESTING FOR BACTERICIDE PROPERTIES OF SILVER NPS

For the test organisms, the following microscopic fungi and bacteria were used: *Bacillus mycoides, Micrococcus flavus; E. coli; Aspergillus niger; Penicillium chrysogenum.*

For the biocide, the growth inhibitor of cultures, the modifier MMA "Akvivon-TM," TU 2499-024-87552538-12 was used.

"Akvivon-TM" modifier in concentrations 0.5, 1.0, and 2.0% was injected in agar located in Petri dishes, 20 mL in each. After solidification of agar, holes of 7 mm in diameter were made in it by a drill, where agar block with pure culture were placed.

Pure cultures of microorganisms were cultivated on a solid agar medium MPA, Saburo (day old cultures of bacteria and three day old cultures of fungi were used).

The test was repeated three times. Every day during 6 days, diameter of the microorganisms growth zone was measured. For the control, growth of cultures on the agar medium without "Akvivon-TM" modifier was taken.

S-shaped kinetic curves of bacteria and microscopic fungi growth are shown in Figs. 1–5.

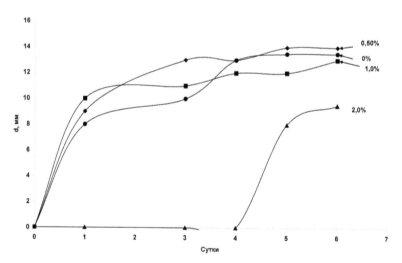

FIGURE 17.1   *E. coli* growth in the agarized medium containing different concentrations of "Akvivon-TM" MMA.

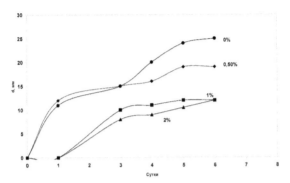

FIGURE 17.2 *Bacillus mycoides* growth in the agarized medium containing different concentrations of "Akvivon-TM" MMA.

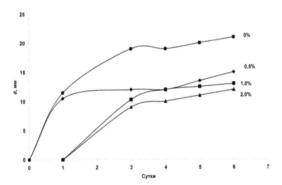

FIGURE 17.3 *Micrococcus flavus* growth in the agarized medium containing different concentrations of "Akvivon-TM" MMA.

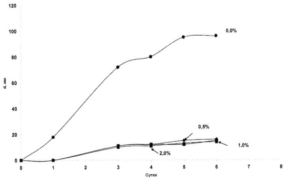

FIGURE 17.4 *Aspergillus niger* growth in the agarized medium containing different concentrations of "Akvivon-TM" MMA.

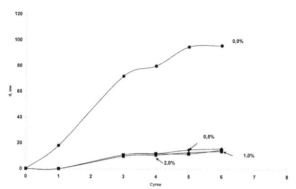

FIGURE 17.5 *Penicillum chrysogenum* growth in the agarized medium containing different concentrations of "Akvivon-TM" MMA.

Previously, using a logistic function [33], the method of kinetic curve analysis for microbial culture growth concluded in determination of the growth rate constant K_c in the presence of various concentrations of biocides [34] and its application [35, 36] has been discussed. These constants are virtually independent of the biocide concentration and are calculated from culture growth rates in the presence of different additive concentrations and in the control (without additive). These rates describe the initial and stationary stages of microbial growth, which allow determination of the activity sequence of additive effect on bacteria, and fungi growth.

It is known that the most enzymatic reactions and the effective rate constant of a fungus colony growth in the presence of biocides is described by the following formula:

$$b_i = b_o \cdot K_c / (K_c + C), \qquad (1)$$

where b_i is the effective rate constant of a fungus colony growth in the presence of the biocide; b_o is the effective rate constant of a fungus colony growth in the absence of the biocide; C is the biocide concentration; K_c is a constant quantitatively equal to the biocide concentration, at which $b_i = b_o/2$ and can be used for assessment of biocide activity.

The lower K_c values are, the stronger the biocide effect is. As Fig. 17.6 shows, addition of silver NPs significantly depresses growth of *Aspergillus niger* and *Penicillium chrysogenum* both at the initial and at stationary stages.

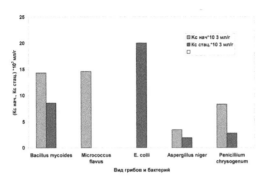

FIGURE 17.6   Histogram of constants for initial and stationary growth of microscopic fungi and bacteria in the presence of silver NPs.

Inhibition of bacterial growth manifests itself in increased induction period (the lag-phase) (Table 17.2).

TABLE 17.2   Induction Periods (days) for Microscopic Fungi and Bacteria Growth in the Presence of Various Silver NP Concentrations

Test organism	Silver NP concentration			
	0%	0.5%	1%	2%
Bacillus mycoides	0	0	1	1
Micrococcus Flavus	0	0	1	1
E. coli	0	0	0	4
Aspergillus niger	0	1	1	1
Penicillium chrysogenum	0	1	1	1

Addition of silver NPs significantly increases the lag-phase. In the case *E. coli* growth (Fig. 17.1), the lag-phase duration reaches four days at the maximum concentration of 2% silver NPs (as bacteria growth inhibitor) in agar, whereas for two bacterial cultures, *Bacillus mycoides* (Fig. 17.2) and *Micrococcus flavus* (Fig. 17.3), the threshold concentration of the inhibitor is 1.0%. For microscopic fungi *Aspergillus niger* (Fig. 17.4) and *Penicillum chrysogenum* (Fig. 17.5), different situation is observed, when growth depression is initiated at 0.5% of silver NPs, and within the first day, equilibrium growth pattern is reached. First and foremost, these results indicate different mechanisms of inhibitor effect on microorganisms. Figure

17.6 shows a histogram of constants for initial and stationary growth of microscopic fungi and bacteria.

As follows from the results obtained, silver NPs depress the growth of microscopic fungi abruptly. In this case, the ultimate value of fungus biomass decreases by 2.5–5 times (Figs. 2 and 3) and not concentration dependence of its increase is observed, whereas for agar, bacterial growth is gradually decelerated with addition of small silver NP concentrations. In the case of *E. coli* growth on agar, the inhibiting effect of silver NPs is only observed in the presence of 2% solution of "Akvivon-TM" MMA.

Apparently, the destructive effect of silver NPs has two mechanisms: electrochemical mechanism associated with accumulation of charges on the cellular membrane surface and free-radical mechanism, when free radicals are released. As is known, cation-containing surfactants intensify the decay of hydroperoxides with free radical formation. Hydroperoxides are the primary products of oxidation of many organic substances by molecular oxygen and are formed spontaneously in materials and products. They are also formed in biochemical processes in living organisms. The cooperative action of cationic surfactants and metal compounds (homogeneous catalysts of hydrocarbon oxidation) is synergistic. Radicals produced by peroxide decay diffuse into cellular membranes and destroy it. If the growth of microscopic fungi is decelerated, the free-radical degradation mechanism is predominant, because no concentration dependence is observed, whereas in case of bacteria the electrochemical mechanism is predominant, and the concentration dependence of bacterial growth depression is observed.

17.4 CONCLUSION

Experimental results are present, showing designing and creation of new generation of modifiers with introduced silver NPs to be applied in advanced technologies, various compositions and materials to adhere them new biological, physicochemical and technical properties. Possible mechanism of silver NP impact on microscopic fungi and bacteria is discussed.

KEYWORDS

- **Biocides**
- **Biotechnology**
- **Fungicides**
- **Nanotechnology**
- **Silver nanoparticles**

REFERENCES

1. Zhao, G., & Jr. Stevens, S. E. (1998). Biometals, 11, 27–32.
2. Furno, F., Morley, K. S., Wong, B., Sharp, B. L., Arnold, P. L., Howdle, S. M. et al. (2004). J. Antimicrob. Chemother, 54, 1019–1024.
3. Jun Sung Kim., Eunye Kuk., KyeongNamYu., Lee, Jong., Hyun Kim, So., Kyung Park, Young.,& Kyung Park, Yong., Cheol-Yong Hwang., Yong-Kwon, Kim., Yoon-Sik, Lee., Dae Hong, Jeong., Myung-Haing, Cho. (2007). Nanomedicine: Nanotechnology, Biology and Medicine, 3(1), 95–101.
4. Aymonier, C., Schlotterbeck, U., Antonietti, L., Zacharias, Ph., Thomann, R., Tiller, J. C., & Mecking, S. (2002). Chem. Commun., 8(24), 3018–3019.
5. Anishchenko, E. V., Lyamina, G. V., Korshikova, N. M., & Mokrousov, G. M. (2006). Izv. TPU, 309, 1.
6. Hamouda, T., Myc, A., Danovan, B., Shih, A., Reuter, J. D., & Jr. Baker, J. R. (2000). Microbiol Res. 156, 1–7.
7. Dibrov, P., Dzioba, J., Gosink, K. K., & Hase, C. C. (2002). Antimicrob. Agents Chemother, 46, 2668–2670.
8. Dragieva, I., Stoeva, S., Stoimenov, P., Pavlikianov, E., & Klabunde, K. (1999). Nanostruct. Mater, 12, 267–270.
9. Sondi, I., Salopek-Sondi, B., & Colloid, J. (2004). Interface. Sci., 275, 177–182.
10. Amro, N. A., Kotra, L. P., Wadu-Mesthrige, K., Bulychev, A., Mobashery, S., & Liu, G. (2000). Langmuir, 16, 2789–2796.
11. Danilczuk, M., Lund, A., Saldo, J., Yamada, H., & Michalik, J. (2006). Spectrochimaca Acta Part A, 63, 189–191.
12. Stoimenov, P. K., Klinger, R. L., Marchin, G. L., & Klabunde, K. J. (2002). Langmuir, 18, 6679.
13. Feng, Q. L., Wu, J., Chen, G. Q., Cui, F. Z., Kim, T. M., Kim, J. O., & Biomed, J. (2000). Mater. Res., 52, 662.
14. Spadaro, J. A., Berger, t. J., Barranco, S. D., Chapin, S. E., & Becker, R. O. (1974). Microb Agents Chemother., 6, 637.
15. Kasaikina, O. T., Kartasheva, Z. S., Pisarenko, L. M., & Obshch. Z. H. (2008). Khim, 78(8), 1298–1309.

16. Kasaikina, O. T., Golyavin, A. A., Krugovov, D. A., Kartasheva, Z. S., & Pisarenko, L. M. (2010). Vestn. MGU, Ser. Khim., 246–250.
17. Mengele, E. A., Kartasheva, Z. S., Plashchina, I. G., Kasaikina, O. T., & Zh, Koll., (2008). 70(6), 805–811.
18. Kasaikina, O. T., Kortenska, V. D., & Kartasheva, Z. S., et al. (1999). Colloid and Surface, A, Physicochemistry and Engineering, 149, 29.
19. Korenevsky, A. A., Sorokin, V. V., & Karavaiko, G. I. (1993). Mikrobiologia, 62(6), 1085–1092
20. Woo Kyung Jung, Hye Cheong Koo., Ki Woo Kim., Sook Shin., So Hyun Kim., & Yong Ho Park, (2008). Appl. and Environmental Microbiology, 74(7), 2171–2178.
21. Zhang, S., & Jr. Crow, S. A., (2001). Applied and Environmental Microbiology, 67(9), 4030–4035.
22. Egorova, E. M., Revina, A. A., Rostovshchikova, T. N., & Kiseleva, O. I., (2001). Vestn. MGU, Ser. 2, Khimia, 42, 332–338.
23. Kumar, A. P., Depan, D., Tomer, N. S., & Singh, R. P. (2009). Progress in polymer science, 34(6), 479–515.
24. Reddy, M. M., Deghton, M., Gupta, R. K., Bhat-acharya, S. N., & Parthasaraty, R. (2009). J. Appl. Polym. Sci., 111, 3, 1426–1432.
25. Gumargalieva, K. Z., Kalinina, I. G., Semenov, S. A., Zaikov, G. E., Zimina, L. A., & Artsis, M. I. (2011). RFP intern., 6, 2, 114–120.
26. Kalinina, I. G., Gumargalieva, K. Z., Kuznetsova, O. N., & Zaikov, G. E. (2012). Vestn. Kazakhsk. Tekhnol. Univ., 15, 12, 115–119.
27. Kalinina, I. G., Belov, G. P., Gumargalieva, K. Z., Petronyuk, Yu. S., & Semenov, S. A. (2011). Khim. Fizika, 30(2), 70–79.
28. Mironova, S. N., Malama, A. A., Filimonova, T. V., Moiseev, Yu. V., Gumargalieva, K. Z., Semenov, S. A., Mironov, V. P., Grushevich, L., & Edokl, A. N. (1985). BSSR, 34(6), 228–560.
29. Gumargalieva, K. Z., & Kalinina, I. G., (2010). Polimernye Materialy, 7–8, 58–62.
30. Gumargalieva, K. Z., Kalinina, I. G., (2010). Polimernye Materialy, 10, 18–24.
31. GOST 9.048–89. ESZKS Technical Products. Methods for Laboratory Tests for Resistance to the Effect of Mold Fungi.
32. GOST 9.049–91. Polymeric Materials and their Components, Methods for Laboratory Tests for Resistance to the Effect of Mold Fungi (Method 3).
33. Emanuel, N. M. (1977). Kinetics of Experimental Tumour Processes, Nauka, Moscow, 354 p.
34. Gumargalieva, K. Z., Kalinina, I. G., Mironova, S. N., & Semenov, S. A. (1988). Mikrobiologia, 57(5), 879–882.
35. Gumargalieva, K. Z., & Zaikov, G. E. (1998). Biodegradation and Biodeterioration of Polymers: Kinetical Aspects. Nova Science Publishers, Inc. Commack, New York, 409 p.
36. Gumargalieva, K. Z., Kalinina, I. G., Zaikov, G. E., Semenov, S. A., & Ryzhkov, A. I. (1996). Chem. Phys. Reports, 15(10), 1463–1476.

CHAPTER 18

TRENDS IN CARBON NANOTUBE/ POLYMER COMPOSITES

A. K. HAGHI and G. E. ZAIKOV

CONTENTS

ABSTRACT

To promote the design and development of CNT-nanocomposite materials, structure and property relationships must be recognized that predict the bulk mechanical of these materials as a function of the molecular and micro structure mechanical properties of nano structured materials can be calculated by a select set of computational methods. In this chapter, new trends in computational chemistry and computational mechanics for the prediction of the structure and properties of CNT materials are presented simultaneously.

18.1   INTRODUCTION

It has been known that the mechanical properties of polymeric materials like stiffness and strength can be engineered by producing composites that are composed of different volume fraction of one or more reinforcing phases. In traditional form, polymeric materials have been reinforced with carbon, glass, basalt, ceramic and aramid microfibers to improve their mechanical properties. These composite materials have been used in many applications in automotive, aerospace and mass transit. As time has proceeded, a practical accomplishment of such composites has begun to change from microscale composites to nanocomposite, taking advantages of better mechanical properties. While some credit can be attributed to the intrinsic properties of the fillers, most of these advantages stem from the extreme reduction in filler size combined with the large enhancement in the specific surface area and interfacial area they present to the matrix phase. In addition, since traditional composites use over 40 wt % of the reinforcing phase, the dispersion of just a few weight percentages of nanofillers into polymeric matrices could lead to dramatic changes in their mechanical properties. One of the earliest nanofiller that witch have received significant and shown super mechanical properties is Carbon Nano Tube (CNT), because of their unique properties, CNTs have a wide range of potentials for engineering applications due to their exceptional mechanical, physical, electrical properties and geometrical characteristics consisting of small diameter and high aspect ratio. It has shown that dispersion of a few weight percentages of nanotubes in a matrix dramatically increase mechanical, thermal and electrical properties of composite materials. Development of CNT-nanocomposites requires a good understanding of CNT's and CNT's nanocomposite properties. Because of the huge cost and technological difficulties associated with experimental analysis at the scale of nano, researchers are encouraged to employ computational methods for simulating the behavior of nanostructures like CNTs from different mechanical points of view.

To promote the design and development of CNT-nanocomposite materials, structure and property relationships must be recognized that predict the bulk mechanical of these materials as a function of the molecular and micro structure mechanical properties of nano structured materials can be calculated by a select set of computational methods. These modeling methods extend across a wide range of length and time scales, as shown in Fig. 18.1. For the smallest length and time scales, a complete understanding of the behavior of materials requires theoretical and computational tools that span the atomic-scale detail of first principles methods (density functional theory, molecular dynamics, and Monte Carlo methods). For the largest length and time scales, computational mechanics is used to predict the mechanical behavior of materials and engineering structures. And the coarser grained description provided by continuum equations. However, the intermediate length and time scales do not have general modeling methods that are as well developed as those on the smallest and largest time and length scales. Therefore, recent efforts have focused on combining traditional methodologies and continuum descriptions within a unified multiscale framework. multiscale modeling techniques are employed, which take advantage of computational chemistry and computational mechanics methods simultaneously for the prediction of the structure and properties of materials.

As illustrated in Fig. 18.1, in each modeling methods has extended classes of related modeling tools that are shown in a short view in a diagram in Fig. 18.2.

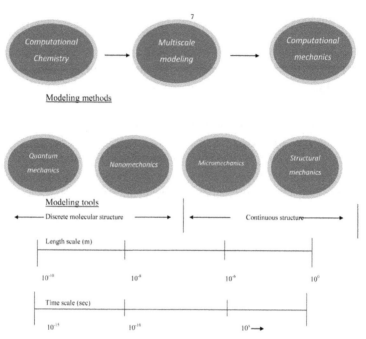

FIGURE 18.1 Different length and time scale used in determination mechanical properties of polymer nano-composite.

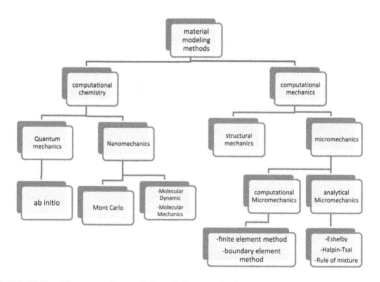

FIGURE 18.2 Diagram of material modeling methods.

18.2 MATERIAL MODELING METHODS

There are different of modeling methods currently used by the researches. They plan not only to simulate material behavior at a particular scale of interest but also to assist in developing new materials with highly desirable properties. These scales can range from the basic atomistic to the much coarser continuum level. The hierarchy of the modeling methods consists quantum mechanics, molecular dynamic, micromechanics and finally continuum mechanics, that could be categorized in tow main group: atomistic modeling and computational mechanics.

18.2.1 COMPUTATIONAL CHEMISTRY (ATOMISTIC MODELING)

The atomistic methods usually employ atoms, molecules or their group and can be classified into three main categories, namely the quantum mechanics (QM), molecular dynamics (MD) and Monte Carlo (MC). Other atomistic modeling techniques such as tight bonding molecular dynamics (TBMD), local density (LD), dissipative particle dynamics (DPD), lattice Boltzmann (LB), Brownian dynamics (BD), time-dependent Ginzburg–Lanau method, Morse potential function model, and modified Morse potential function model were also applied afterwards.

18.2.1.1 QUANTUM MECHANICS

The observable properties of solid materials are governed by quantum mechanics, as expressed by solutions of a Schrödinger equation for the motion of the electrons and the nuclei. However, because of the inherent difficulty of obtaining even coarsely approximate solutions of the full many body Schrödinger equation, one typically focuses on reduced descriptions that are believed to capture the essential energetic of the problem of interest. Two main quantum mechanics method are "ab initio" and "density function method (DFT)."

Unlike most materials simulation methods that are based on classical potentials, the main advantages of ab initio methods, which is based on first principles density functional theory (without any adjust able parameters), are the generality, reliability, and accuracy of these methods.

They involve the solution of Schrödinger's equation for each electron, in the self-consistent potential created by the other electrons and the nuclei. Ab initio methods can be applied to a wide range of systems and properties. However, these techniques are computationally exhaustive, making them difficult for simulations involving large numbers of atoms. There are three widely used procedures in ab initio simulation. These procedures are single point calculations, geometry optimization, and frequency calculation. Single point calculations involve the determination of energy and wave functions for a given geometry. This is often used as a preliminary step in a more detailed simulation. Geometry calculations are used to determine energy and wave functions for an initial geometry, and subsequent geometries with lower energy levels. A number of procedures exist for establishing geometries at each calculation step. Frequency calculations are used to predict Infrared and Raman intensities of a molecular system. Ab initio simulations are restricted to small numbers of atoms because of the intense computational resources that are required ab initio techniques have been used on a limited basis for the prediction of mechanical properties of polymer based nanostructured composites.

18.2.1.2   MOLECULAR DYNAMIC

MD is a simulation technique that used to estimate the time depended physical properties of a system of interacting particles (e.g., atoms, molecules, etc.) by predict the time evolution. MD simulation used to investigating the structure, dynamics, and thermodynamics of individual molecules.

There are two basic assumptions made in standard molecular dynamics simulations. Molecules or atoms are described as a system of interacting material points, whose motion is described dynamically with a vector of instantaneous positions and velocities. The atomic interaction has a strong dependence on the spatial orientation and distances between separate atoms. This model is often referred to as the soft sphere model, where the softness is analogous to the electron clouds of atoms.

No mass changes in the system. Equivalently, the number of atoms in the system remains the same.

The atomic position, velocities, and accelerations of individual particles that vary with time The described by track that MD simulation generates and then used to obtain average value of system such as energy, pressure and temperature.

The main three parts of MD simulation are:

- initial conditions (e.g., initial positions and velocities of all particles in the system);
- The interaction potentials between particles to represent the forces among all the particles;
- The evolution of the system in time by solving of classical Newtonian equations of motion for all particles in the system.

The equation of motion is generally given by Eq. (1).

$$\vec{F}_i(t) = m_i \frac{d^2 \vec{r}_i}{dt^2} \tag{1}$$

where $\vec{F}_i(t)$ is the force acting on the i-th atom or particle at time t and is obtained as the negative gradient of the interaction potential U, m_i is the atomic mass and $\vec{r}_i$ the atomic position. The interaction potentials together with their parameters, describe how the particles in a system interact with each other (so-called force field). Force field may be obtained by quantum method (e.g., ab initio), empirical method (e.g., Lennard–Jones, Mores, and Born-Mayer) or quantum-empirical method (e.g., embedded atom model, glue model, bond order potential).

The criteria for selecting a force field include the accuracy, transferability and computational speed. A typical interaction potential U may consist of a number of bonded and nonbonded interaction terms:

$$U(\vec{r_1}, \vec{r_2}, \vec{r_3}, ..., \vec{r_n})$$

$$= \sum_{i_{bond}}^{N_{bond}} U_{bond}(i_{bond}, \vec{r_a}, \vec{r_b}) +$$

$$\sum_{i_{angle}}^{N_{angle}} U_{angle}(i_{angle}, \vec{r_a}, \vec{r_b}, \vec{r_c}) + \sum_{i_{torsion}}^{N_{torsion}} U_{torsion}(i_{torsion}, \vec{r_a}, \vec{r_b}, \vec{r_c}, \vec{r_d})$$

$$+ \sum_{i_{inversion}}^{N_{inversion}} U_{inversion}(i_{inversion}, \vec{r_a}, \vec{r_b}, \vec{r_c}, \vec{r_d})$$

$$+ \sum_{i=1}^{N-1} \sum_{j>i}^{N} U_{vdw}(i, j, \vec{r_a}, \vec{r_b})$$

$$+ \sum_{i=1}^{N-1} \sum_{j>i}^{N} U_{electrostatic}(i, j, \vec{r_a}, \vec{r_b}) \tag{2}$$

The first four terms represent bonded interactions, i.e., bond stretching U_{bond}, bond-angle bend U_{angle}, dihedral angle torsion $U_{torsion}$ and inversion interaction $U_{inversion}$. Vander Waals energy U_{vdw} and electrostatic energy $U_{electrostatic}$ are non-bonded interactions. In the equation, $\vec{r}_a, \vec{r}_b, \vec{r}_c, \vec{r}_d$ are the positions of the atoms or particles specifically involved in a given interaction; N_{bond}, N_{angle}, $N_{torsion}$ and $N_{inversion}$ illustrate the total numbers of interactions in the simulated system; i_{bond}, i_{angle}, $i_{torsion}$ and $i_{inversion}$ presented an individual interaction each of them. There are many algorithms like *varlet, velocity varlet, leap-frog* and *Beeman*, for integrating the equation of motion, all of them using finite difference methods, and assume that the atomic position $\vec{r}_i$, velocities $\vec{v}$ and accelerations $\vec{a}$ can be approximated by a *Taylor series expansion*:

$$\vec{r}(t + \delta t) = \vec{r}(t) + \vec{v}(t)\delta t + \frac{1}{2}\vec{a}(t)\delta^2 t + \cdots \tag{3}$$

$$\vec{v}(t + \delta t) = \vec{v}(t) + \vec{a}(t)\delta t + \frac{1}{2}\vec{b}(t)\delta^2 t + \cdots \tag{4}$$

$$\vec{a}(t + \delta t) = \vec{a}(t) + \vec{b}(t)\delta t + \cdots \tag{5}$$

The *varlet algorithm* is probably the most widely used method. It uses the positions $\vec{r}(t)$ and accelerations $\vec{a}(t)$ at time t, and the positions $\vec{r}(t+\delta t)$ from the previous step $(t - \delta)$ to calculate the new positions $\vec{r}(t + \delta t)$ at $(t + \delta t)$, so:

$$\vec{r}(t + \delta t) = \vec{r}(t) + \vec{v}(t)\delta t + \frac{1}{2}\vec{a}(t)\delta t^2 + \cdots \tag{6}$$

$$\vec{r}(t - \delta t) = \vec{r}(t) - \vec{v}(t)\delta t + \frac{1}{2}\vec{a}(t)\delta t^2 + \cdots \tag{7}$$

$$\vec{r}(t + \delta t) = 2\vec{r}(t) - \vec{r}(t - \delta t) + \vec{a}(t)\delta t^2 + \cdots \tag{8}$$

The velocities at time t and $t + 1/2\delta t$ can be respectively estimated.

$$\vec{v}(t) = [\vec{r}(t + \delta t) - \vec{r}(t - \delta t)]/2\delta t \tag{9}$$

$$\vec{v}(t + 1/2\delta t) = [\vec{r}(t + \delta t) - \vec{r}(t - \delta t)]/\delta t \tag{10}$$

The advantages of this method are *time-reversible* and *good energy conservation* properties, where disadvantage is *low memory storage* because the velocities are not included in the time integration. However, removing the velocity from the integration introduces numerical inaccuracies method, namely the *velocity Verlet* and *Verlet leap-frog algorithms* as mentioned before, which clearly involve the velocity in the time evolution of the atomic coordinates.

18.2.1.3 MONTE CARLO

Monte Carlo technique (*Metropolis method*) use random number from a given probability distribution to generate a sample population of the system from which one can calculate the properties of interest. a MC simulation usually consists of three typical steps:
- The physical problem is translated into an analogous probabilistic or statistical model.
- The probabilistic model is solved by a numerical stochastic sampling experiment.
- The obtained data are analyzed by using statistical methods.

Athwart MD, which provides information for nonequilibrium as well as equilibrium properties, MC gives only the information on equilibrium properties (e.g., free energy, phase equilibrium). In a NVT ensemble with N atoms, new formation the change in the system *Hamiltonian* (ΔH) by randomly or systematically moving one atom from position $i \rightarrow j$ can calculated.

$$\Delta H = H(j) - H(i) \tag{11}$$

where $H(i)$ and $H(j)$ are the *Hamiltonian* associated with the original and new configuration, respectively.

The $\Delta H < 0$ shows the state of lower energy for system. So, the movement is accepted and the new position is stable place for atom.

For $\Delta H \geq 0$, the move to new position is accepted only with a certain probability $Pi \rightarrow j$ which is given by

$$Pi \rightarrow j \propto exp\left(-\frac{\Delta H}{K_B T}\right) \tag{12}$$

K_B is the Boltzmann constant.

According to *Metropolis* et al. a random number ζ between 0 and 1 could be generated and determine the new configuration according to the following rule:

$$\text{For } \zeta \leq exp\left(-\frac{\Delta H}{K_B T}\right); \text{ the move is accepted;} \tag{13}$$

$$\text{For } \zeta > exp\left(-\frac{\Delta H}{K_B T}\right); \text{ the move is not accepted.} \tag{14}$$

If the new configuration is rejected, repeats the process by using other random chosen atoms.

In a μVT ensemble, a new configuration j by arbitrarily chosen and it can be exchanged by an atom of a different kind. This method affects the chemical composition of the system and the move is accepted with a certain probability. However, the energy, ΔU, will be changed by change in composition.

If $\Delta U < 0$, the move of compositional change is accepted. However, if $\Delta U \geq 0$, the move is accepted with a certain probability which is given by:

$$Pi \rightarrow j \propto exp\left(-\frac{\Delta U}{K_B T}\right) \tag{15}$$

where ΔU is the change in the sum of the mixing energy and the chemical potential of the mixture. If the new configuration is rejected one counts the original configuration as a new one and repeats the process by using some other arbitrarily or systematically chosen atoms. In polymer nanocomposites, MC methods have been used to investigate the molecular structure at nanoparticle surface and evaluate the effects of various factors.

18.2.2 COMPUTATIONAL MECHANICS

The continuum material that assumes is continuously distributed through-out its volume have an average density and can be subjected to body forces such as gravity and surface forces. Observed macroscopic behavior is usually illustrated by ignoring the atomic and molecular structure. The basic laws for continuum model are:

- continuity, drawn from the conservation of mass.
- equilibrium, drawn from momentum considerations and Newton's second law.
- the moment of momentum principle, based on the model that the time rate of change of angular momentum with respect to an arbitrary point is equal to the resultant moment.
- conservation of energy, based on the first law of thermodynamics.
- Conservation of entropy, based on the second law of thermodynamics.

These laws provide the basis for the continuum model and must be coupled with the appropriate constitutive equations and the equations of state to provide all the equations necessary for solving a continuum problem. The continuum method relates the deformation of a continuous medium to the external forces acting on the medium and the resulting internal stress and strain. Computational approaches range from simple closed-form analytical expressions to micromechanics and complex structural mechanics calculations based on beam and shell theory. In this section, we introduce some continuum methods that have been used in polymer nano-composites, including micromechanics models (e.g., Halpin–Tsai model, Mori–Tanaka model and finite element analysis.) and the semicontinuum methods like equivalent-continuum model and will be discussed in the next section.

18.2.2.1 MICROMECHANICS

Micromechanics are a study of mechanical properties of unidirectional composites in terms of those of constituent materials. In particular, the properties to be discussed are elastic modulus, hydrothermal expansion coefficients and strengths. In discussing composites properties it is important to define a *volume element,* which is small enough to show the microscopic structural details, yet large enough to present the overall behavior of the composite. Such a volume element is called the *Representative Volume Element (RVE)*. A simple representative volume element can consists of a fiber embedded in a matrix block, as shown in Fig. 18.3.

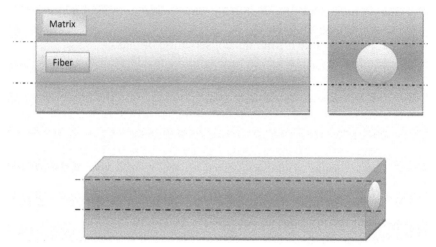

FIGURE 18.3   A Representative Volume Element (RVE). The total volume and mass of each constituent are denoted by V and M, respectively. The subscripts m and f stand for matrix and fiber, respectively.

One a representative volume element is chosen, proper boundary conditions are prescribed. Ideally, these boundary conditions must represent the in situ state of stress and strain within the composite. That is, the prescribed boundary conditions must be the same as those if the representative volume element were actually in the composite. Finally, a prediction of composite properties follows from the solution of the foregoing boundary value problem. Although the procedure involved is conceptually simple, the actual solution is rather difficult. Consequently, many assumption and approximation have been introduces, and therefore various solution are available.

18.2.2.1.1   BASIC CONCEPTS

Micromechanics models usually used to reinforced polymer composites, based on follow basic assumptions:

1. Linear elasticity of fillers and polymer matrix.
2. The reinforcement are axis-symmetric, identical in shape and size, and can be characterized by parameters such as length-to-diameter ratio (aspect ratio).
3. Perfect bonding between reinforcement and polymer interface and the ignorance of interfacial slip, reinforcement and polymer debonding or matrix cracking.

Consider a composite of mass M and volume V, illustrated schematically in Fig. 18.3, V is the volume of a *Representative Volume Element (RVE)*, since the composite is made of fibers and matrix, the mass M is the sum of the total mass M_f of fibers and mass M_m of matrix:

$$M = M_f + M_m \tag{16}$$

Equation (16) is valid regardless of voids, which may be present. However, the composite volume V includes the volume V_v of voids so that:

$$V = V_f + V_m + V_v \tag{17}$$

Dividing Eqs. (16) and (17), leads to the following relation for the mass fraction and volume fractions:

$$m_f + m_m = 1 \tag{18}$$

$$v_f + v_m + v_v = 1 \tag{19}$$

The composite density ρ calculated as follows:

$$\rho = \frac{M}{V} = \frac{(\rho_f v_f + \rho_m v_m)}{V} = \rho_f v_f + \rho_m v_m \tag{20}$$

$$\rho = \frac{1}{m_f / \rho_f + m_m / \rho_m + v_v / \rho} \tag{21}$$

These equations can be used to determine the void fraction:

$$v_v = 1 - \rho(\frac{m_f}{\rho_f} + \frac{m_m}{\rho_m}) \tag{22}$$

The mass fraction of fibers can be measured by removing the matrix. Based on the first concept, the linear elasticity, the linear relationship between the total stress and infinitesimal strain tensors for the reinforcement and matrix as expressed by the following constitutive equations:

$$\sigma_f = C_f \, \varepsilon_f \tag{23}$$

$$\sigma_m = C_m \, \varepsilon_m \tag{24}$$

where C is the stiffness tensor.

The second concept is the average stress and strain. While the stress field σ_i and the corresponding strain field ε_i are usually non-uniform in polymer composites, the average stress $\bar{\sigma}_i$ and strain $\bar{\varepsilon}_i$ are then defined over the representative averaging volume V, respectively. Hypothesize the stress field in the RVE is σ_i, then composite stress $\bar{\sigma}_i$ and is defined by:

$$\bar{\sigma}_i = \frac{1}{V}\int \sigma_i dv = \frac{1}{V}[\int_{v_f} \sigma_i dv + \int_{v_m} \sigma_i dv + \int_{v_v} \sigma_i dv] \tag{25}$$

$$\bar{\sigma}_{fi} = \frac{1}{V_f}\int_{v_f} \sigma_i dV, \bar{\sigma}_{mi} = \frac{1}{V_m}\int_{v_m} \sigma_i dV \tag{26}$$

Because of no stress is transmitted in the voids, $\sigma_i = 0$ in V_v and so:

$$\bar{\sigma}_i = v_f \bar{\sigma}_{fi} + v_m \bar{\sigma}_{mi} \tag{27}$$

where $\vec{\sigma}_i$ is composite average stress, $\bar{\sigma}_{fi}$ is fibers average stress and $\bar{\sigma}_{mi}$ is matrix average stress. Similarly to the composite stress, the composite strain is defined as the volume average strain, and is obtained as:

$$\bar{\varepsilon}_i = v_f \bar{\varepsilon}_{fi} + v_m \bar{\varepsilon}_{mi} + v_v \bar{\varepsilon}_{vi} \tag{28}$$

Despite the stress, the void in strain does not vanished; it is defined in term of the boundary displacements of the voids. So, because of the void fraction is usually negligible. Therefore last term in Eq. (26) could be neglected and the equation corrected to:

$$\overline{\varepsilon}_i = v_f \overline{\varepsilon}_{fi} + v_m \overline{\varepsilon}_{mi} \tag{29}$$

The average *stiffness* of the composite is the tensor C that related the average strain to the average stress as follow equation:

$$\overline{\sigma} = C \overline{\varepsilon} \tag{30}$$

The average *compliance S* is defined in this way:

$$\overline{\varepsilon} = S \overline{\sigma} \tag{31}$$

Another important concept is the *strain concentration* and *stress concentration* tensors A and B, which are basically the ratios between the average reinforcement strain or stress and the corresponding average of the composites.

$$\overline{\varepsilon}_f = A\overline{\varepsilon} \tag{32}$$

$$\overline{\sigma}_f = B\overline{\sigma} \tag{33}$$

Finally, the average composite stiffness can be calculated from the strain concentration tensor A and the reinforcement and matrix properties:

$$C = C_m + v_f (C_f - C_m)A \tag{34}$$

18.2.2.1.2   HALPIN–TSAI MODEL

Halpin-Tsai theory is used for prediction elastic modulus of unidirectional composites as function of aspect ratio. The longitudinal stiffness, E_{11} and transverse modulus, E_{22}, are expressed in the following general form:

$$\frac{E}{E_m} = \frac{1 + \zeta \eta v_f}{1 - \eta v_f} \tag{35}$$

where E and E_m are modulus of composite and matrix respectively, v_f is fiber volume fraction and η is given by this equation:

$$\eta = \frac{\dfrac{E}{E_m} - 1}{\dfrac{E_f}{E_m} + \zeta_f} \tag{36}$$

where η fiber modulus is E_f and ζ is shape parameter that depended on reinforcement geometry and loading direction. For E_{11} calculation, ζ is equal to l/t where l is length and t is thickness of reinforcement, for E_{22}, ζ is equal to w/t where w is width of reinforcement.

For $\zeta_f \rightarrow 0$, the Halpin-Tsai theory converged to inversed *rule of mixture* for stiffness.

$$\frac{1}{E} = \frac{v_f}{E_f} + \frac{1 - v_f}{E_m} \tag{37}$$

If $\zeta_f \rightarrow \infty$, the Halpin-Tsai converge to rule of mixture.

$$E = E_f v_f + E_m (1 - v_f) \tag{38}$$

18.2.2.1.3   MORI–TANAKA MODEL

The Mori-Tanaka model is uses for prediction an elastic stress field for in and around an ellipsoidal reinforcement in an infinite matrix. This method is based on Eshelby's model. Longitudinal and transverse elastic modulus, E_{11} and E_{22}, for isotropic matrix and directed spherical reinforcement are:

$$\frac{E_{11}}{E_m} = \frac{A_0}{A_0 + v_f (A_1 + 2v_0 A_2)} \tag{39}$$

$$\frac{E_{22}}{E_m} = \frac{2A_0}{2A_0 + v_f(-2A_3 + (1-v_0A_4)+(1+v_0)A_5A_0)} \tag{40}$$

where E_m is the modulus of the matrix, vf is the volume fraction of reinforcement, v_0 is the Poisson's ratio of the matrix, parameters, $A_0, A_1, ..., A_5$ are functions of the Eshelby's tensor.

18.2.2.2   FINITE ELEMENT METHODS

The traditional framework in mechanics has always been the continuum. Under this framework, materials are assumed to be composed of a divisible continuous medium, with a constitutive relation that remains the same for a wide range of system sizes. Continuum equations are typically in the form of deterministic or stochastic partial differential equation (FDE's). The underling atomic structure of matter is neglected altogether and is replaced with a continuous and differentiable mass density. Similar replacement is made for other physical quantities such as energy and momentum. Differential equations are then formulated from basic physical principles, such as the conservation of energy or momentum. There are a large variety of numerical method that can be used for solving continuum partial differential equation, the most popular being the finite element method (FEM).

The finite element method is a numerical method for approximating a solution to a system of partial differential equations the FEM proceeds by dividing the continuum into a number of elements, each connected to the next by nodes. This discretization process converts the PDE's into a set of coupled ordinary equations that are solved at the nodes of the FE mesh and interpolated throughout the interior of the elements using shape functions. The main advantages of the FEM are its flexibility in geometry, refinement, and loading conditions. It should be noted that the FEM is local, which means that the energy within a body does not change throughout each element and only depends on the energy of the nodes of that element.

The total potential energy under the FE framework consists of two parts, the internal energy U and external work W.

$$\Pi = U - W \tag{41}$$

The internal energy is the strain energy caused by deformation of the body and can be written as:

$$U = \frac{1}{2}\int_{\Omega}\{\sigma\}^T\{\varepsilon\}d\Omega = \frac{1}{2}\int_{\Omega}\{\sigma\}^T[D]\{\varepsilon\}d\Omega \tag{42}$$

where $\{\sigma\} = \{\{\sigma\} = \{\sigma_{xx}\sigma_{yy}\sigma_{zz}\tau_{xy}\tau_{yz}\tau_{xz}\}^T\}^T$ denotes the stress vector, and $\{\varepsilon\} = \{\{\varepsilon_{xx}\varepsilon_{yy}\varepsilon_{zz}\gamma_{xy}\gamma_{yz}\gamma_{xz}\}^T\}^T$ denotes the strain vector, [D] is the elasticity matrix, and Ω indicate that integration is over the entire domain.

The external work can be written as:

$$W = \int_{\Omega}\{u\,v\,w\}\begin{bmatrix}\mathfrak{I}_x\\\mathfrak{I}_y\\\mathfrak{I}_z\end{bmatrix}d\Omega + \int_{\Gamma}\{u\,v\,w\}\begin{bmatrix}T_x\\T_y\\T_z\end{bmatrix}d\Gamma \tag{43}$$

where u, v and w represent the displacement in x, y, z directions, respectively, $\{\mathfrak{I}\}$ is the force vector which contain both applied and body forces, $\{T\}$ is the surface traction vector, and Γ indicates that the integration of the traction occurs only over the surface of the body. After discretization of the region into a number of elements, point-wise discretization of the displacements u, v and z directions, is achieved using shape function [N] for each element, such that the total potential energy becomes:

$$\Pi^g = \frac{1}{2}\{d\}^T\int_{\Omega^g}[B]^T[D][B]\{d\}d\Omega - \{d\}^T\int_{\Omega^g}[N]^T\begin{bmatrix}\mathfrak{I}_x\\\mathfrak{I}_y\\\mathfrak{I}_z\end{bmatrix}d\Omega - \{d\}^T\int_{\Gamma^g}[N]^T\begin{bmatrix}T_x\\T_y\\T_z\end{bmatrix}d\Gamma \tag{44}$$

where [B] is the matrix containing the derivation of the shape function, and $\{d\}$ is a vector containing the displacements.

18.3   CARBON NANO TUBES (CNT): STRUCTURE AND PROPERTIES

A Carbon Nanotube is a tube shaped material, made of carbon, having length-to-diameter ratio over of 100,000,000:1, considerably higher than for any other material. These cylindrical carbon molecules have strange properties such as extraordinary mechanical, electrical properties and thermal conductivity, which are important for different fields of materials science and technology.

Nanotubes are members of the globular shape structural category. They are formed from rolling of one atom thick sheets of carbon, called graphene, which cause to hollow structure. The process of rolling could be done at specific and discrete chiral angles, as nanotube properties are dependent to the rolling angle. The single nanotubes physically line up themselves into ropes situation together by van der Waals forces.

Chemical bonding in nanotubes describes by orbital hybridization. The chemical bonding of nanotubes is constituted completely of sp^2 bonds, similar to those of graphite, which are stronger than the sp^3 bonds found in diamond, provide nanotubes with their unique strength.

18.3.1 CATEGORIZATION OF NANOTUBES

Carbon nanotube is classified as single walled nanotubes (SWNTs), multiwalled nanotubes (MWNTs) and Double walled carbon nanotubes (DWNTs).

18.3.1.1 SINGLE-WALLED CARBON NANOTUBE

SWNTs have a diameter of near to 1nanometer and tube length of longer than 10^9 times. The structure of a SWNT can be explained by wresting a graphene into a seamless cylinder. The way the graphene is wrested is depicted by a pair of indices *(n,m)*. The integer's *n* and *m* indicate the number of unit vectors in the direction of two points in the honeycomb crystal lattice of graphene. If *n = m*, the nanotubes are called *armchair* nanotubes and while *m= 0*, the nanotubes are called *zigzag* nanotubes (Fig. 18.4). If not, they are named *chiral*. The *diameter* of a perfect nanotube can be calculated from its *(n,m)* indices as:

$$d = \frac{a}{\delta}\sqrt{\left(n^2 + nm + m^2\right)} = 78.3\sqrt{\left((n+m)^2 - nm\right)}\,(pm) \quad (a = 0.246 \text{ nm}) \quad (45)$$

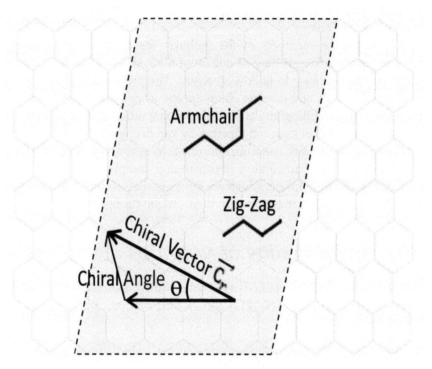

FIGURE 18.4   Schematic diagram of a hexagonal grapheme sheet.

SWNTs are an important kind of carbon nanotube due to most of their properties change considerably with the *(n,m)* values, and according to *Kataura plot*, this dependence is unsteady. Mechanical properties of single SWNTs were predicted remarkable by Quantum mechanics calculations as Young's modulus of 0.64–1 TPa, Tensile Strength of 150–180 GPa, strain to failure of 5–30% while having a relatively low density of 1.4–1.6 g/cm^3.

These high stiffness and superior mechanical properties for SWNTs are due to the chemical structure of the repeat unit. The repeat unit is composed completely of sp^2-hybridized carbons and without any points for flexibility or rotation. Also, Single walled nanotubes with diameters of an order of a nanometer can be excellent conductors for electrical industries.

In most cases, SWNTs are synthesized through the reaction of a gaseous carbon feedstock to form the nanotubes on catalyst particles.

18.3.1.2 MULTI-WALLED CARBON NANOTUBE

Multi-walled nanotubes (MWNT) consist of several concentric graphene tubes which its interlayer distance is around 3.4 Å; close to the distance between graphene layers in graphite. Its single shells in MWNTs can be explained as SWNTs.

18.3.1.3 DOUBLE-WALLED CARBON NANOTUBES (DWNT)

Double-walled carbon nanotubes consist of a particular set of nanotubes because their morphology and properties are comparable to those of SWNT but they have considerably superior chemical resistance. Prominently, it is important when grafting of chemical functions at the surface of the nano-tubes is required to achieve new properties of the CNT. In this processing of SWNT, some $C=C$ double bonds will be broken by covalent function-alization, and thus, both mechanical and electrical properties of nanotubes will be modified. About of DWNT, only the outer shell is modified.

18.3.2 CNT PROPERTIES

18.3.2.1 STRENGTH

In terms of strength, tensile strength and elastic modulus are explained and it has not been discovered any material as strong as carbon nanotubes yet. CNTs due to containing the single carbon atoms, having the covalent sp^2 bonds formed between them and they could resist against high tensile stress. Many studies have been done on tensile strength of carbon nano-tubes and totally it was included that single CNT shells have strengths of about 100 GPa. Since density of a solid carbon nanotubes is around of 1.3 to 1.4 g/cm^3, specific strength of them is up to 48000 kN·m·kg^{-1} which it causes to including carbon nanotube as the best of known materials, compared to high carbon steel that has specific strength of 154 kN·m·kg^{-1}.

Under extreme tensile strain, the tubes will endure plastic deformation, which means the deformation is invariable. This deformation commences at strains of around 5% and can enhance the maximum strain the tubes undergo before breakage by releasing strain energy.

Despite of the most high strength of single CNT shells, weak shear interactions between near shells and tubes leads to considerable diminutions in the effective strength of multiwalled carbon nanotubes, while crosslink in inner shells and tubes, included the strength of these materials is about 60 GPa for multiwalled carbon nanotubes and about 17 GPa for double-walled carbon nanotube bundles.

Almost, hollow structure and high aspect ratio of carbon nanotubes lead to their tendency to suffer bending when placed under compressive, torsion stress.

18.3.2.2   HARDNESS

The regular single-walled carbon nanotubes have ability to undergo a transformation to great hard phase and so, they can endure a pressure up to 25 GPa without deformation. The bulk modulus of great hard phase nanotubes is around 500 GPa, which is higher than that of diamond (420 GPa for single diamond crystal).

18.3.2.3   ELECTRICAL PROPERTIES

Because of the regularity and exceptional electronic structure of graphene, the structure of a nanotube affects its electrical properties strongly. It has been concluded that for a given (n,m) nanotube, while $n=m$, the nanotube is metallic; if $n-m$ is a multiple of 3, then the nanotube is semiconducting with a very small band gap, if not, the nanotube is a moderate semiconductor. However, some exceptions are in this rule, because electrical properties can be strongly affected by curvature in small diameter carbon nanotubes. In theory, metallic nanotubes can transmit an electric current density of 4×109 A/cm^2, which is more than 1,000 times larger than those of metals such as copper, while electro migration lead to limitation of current densities for copper interconnects.

Because of nanoscale cross-section in carbon nanotubes, electrons spread only along the tube's axis and electron transfer includes quantum effects. As a result, carbon nanotubes are commonly referred to as one-dimensional conductors. The maximum electrical transmission of a SWNT is $2G_0$, where $G_0 = 2e^2/h$ is the transmission of a single ballistic quantum channel.

18.3.2.4   THERMAL PROPERTIES

It is expected that nanotubes act as very good thermal conductors, but they are good insulators laterally to the tube axis. Measurements indicate that SWNTs have a room temperature thermal conductivity along its axis of about 3500 $W \cdot m^{-1} \cdot K^{-1}$; higher than that for copper (385 $W \cdot m^{-1} \cdot K^{-1}$). Also, SWNT has a room temperature thermal conductivity across its axis (in the radial direction) of around 1.52 $W \cdot m^{-1} \cdot K^{-1}$, which is nearly as thermally conductive as soil. The temperature constancy of carbon nanotubes is expected to be up to 2800 °C in vacuum and about 750 °C in air.

18.3.2.5   DEFECTS

As with any material, the essence of a crystallographic defect affects the material properties. Defects can happen in the form of atomic vacancies. High levels of such defects can drop the tensile strength up to 85%. A main example is the Stone Wales defect, which makes a pentagon and heptagon pair by reorganization of the bonds. Having small structure in carbon nanotubes lead to dependency of their tensile strength to the weakest segment.

Also, electrical properties of CNTs can be affected by crystallographic defects. A common result is dropped conductivity through the defective section of the tube. A defect in conductive nanotubes can cause the adjacent section to become semiconducting, and particular monatomic vacancies induce magnetic properties.

Crystallographic defects intensively affect the tube's thermal properties. Such defects cause to phonon scattering, which in turn enhance the relaxation rate of the phonons. This decreases the mean free path and declines the thermal conductivity of nanotube structures. Phonon transport simulations show that alternative defects such as nitrogen or boron will mainly cause to scattering of high frequency optical phonons. However, larger scale defects such as Stone Wales defects lead to phonon scattering over a wide range of frequencies, causing to a greater diminution in thermal conductivity.

18.3.3. METHODS OF CNT PRODUCTION

18.3.3.1 ARC DISCHARGE METHOD

Nanotubes were perceived in 1991 in the carbon soot of graphite electrodes during an arc discharge, by using a current of 100 amps that was intended to create fullerenes. However, for the first time, macroscopic production of carbon nanotubes was done in 1992 by the similar method of 1991. During this process, the carbon included the negative electrode sublimates due to the high discharge temperatures. As the nanotubes were initially discovered using this technique, it has been the most widely used method for synthesis of CNTs. The revenue for this method is up to 30% by weight and it produces both single and multi walled nanotubes with lengths of up to 50 micrometers with few structural defects.

18.3.3.2 LASER ABLATION METHOD

Laser ablation method was developed by Dr. Richard Smalley and coworkers at Rice University. In that time, they were blasting metals with a laser to produce a variety of metal molecules. When they noticed the existence of nanotubes they substituted the metals with graphite to produce multiwalled carbon nanotubes. In the next year, the team applied a composite of graphite and metal catalyst particles to synthesize single walled carbon nanotubes. In laser ablation method, vaporizing a graphite target in a high temperature reactor is done by a pulsed laser while an inert gas is bled into the chamber. Nanotubes expand on the cooler surfaces of the reactor as the vaporized carbon condenses. A water cooled surface may be contained in the system to gathering the nanotubes.

The laser ablation method revenues around 70% and produces principally single-walled carbon nanotubes with a controllable diameter determined by the reaction temperature. However, it is more costly than either arc discharge or chemical vapor deposition.

18.3.3.3 PLASMA TORCH

In 2005, a research group from the University of Sherbrooke and the National Research Council of Canada could synthesize Single walled carbon nanotubes by the induction thermal plasma method. This method is alike to arc discharge in that both apply ionized gas to achieve the high temperature necessary to vaporize carbon containing substances and the metal catalysts necessary for the following nanotube development. The thermal plasma is induced by high frequency fluctuating currents in a coil, and is kept in flowing inert gas. Usually, a feedstock of carbon black and metal catalyst particles is supplied into the plasma, and then cooled down to constitute single walled carbon nanotubes. Various single wall carbon nanotube diameter distributions can be synthesized.

The induction thermal plasma method can create up to 2 grams of nanotube material per minute, which is higher than the arc-discharge or the laser ablation methods.

18.3.3.4   CHEMICAL VAPOR DEPOSITION (CVD)

In 1952 and 1959, the catalytic vapor phase deposition of carbon was studied, and finally, in1993; the carbon nanotubes were constituted by this process. In 2007, researchers at the University of Cincinnati (UC) developed a process to develop aligned carbon nanotube arrays of 18 mm length on a First Nano ET3000 carbon nanotube growth system.

In CVD method, a substrate is prepared with a layer of metal catalyst particles, most usually iron, cobalt, nickel or a combination. The metal nanoparticles can also be formed by other ways, including reduction of oxides or oxides solid solutions. The diameters of the carbon nanotubes, which are to be grown, are related to the size of the metal particles. This can be restrained by patterned (or masked) deposition of the metal, annealing, or by plasma etching of a metal layer. The substrate is heated to around of 700 °C. To begin the enlargement of nanotubes, two types of gas are bled into the reactor: a process gas (such as ammonia, nitrogen or hydrogen) and a gas containing carbon (such as acetylene, ethylene, ethanol or methane). Nanotubes develop at the sites of the metal catalyst; the gas containing carbon is broken apart at the surface of the catalyst particle, and the carbon is transferred to the edges of the particle, where it forms the nanotubes. The catalyst particles can stay at the tips of the growing nanotube during expansion, or remain at the nanotube base, depending on the adhesion between the catalyst particle and the substrate.

CVD is a general method for the commercial production of carbon nanotubes. For this idea, the metal nanoparticles are mixed with a catalyst support such as MgO or Al_2O_3 to enhance the surface area for higher revenue of the catalytic reaction of the carbon feedstock with the metal particles. One matter in this synthesis method is the removal of the catalyst support via an acid treatment, which sometimes could destroy the primary structure of the carbon nanotubes. However, other catalyst supports that are soluble in water have verified effective for nanotube development of the different means for nanotube synthesis, CVD indicates the most promise for industrial scale deposition, due to its price/unit ratio, and because CVD is capable of increasing nanotubes directly on a desired substrate, whereas the nanotubes must be collected in the other expansion techniques. The development sites are manageable by careful deposition of the catalyst.

18.3.3.5   SUPER-GROWTH CVD

Kenji Hata, Sumio Iijima and co-workers at AIST, Japan; developed super growth CVD (water assisted chemical vapor deposition), by adding water into CVD reactor to improve the activity and lifetime of the catalyst. Dense millimeter tall nanotube *forests*, aligned normal to the substrate, were created. The *forests* expansion rate could be extracted, as:

$$H(t) = \beta\tau_0(1 - e^{-\frac{t}{\tau_0}}) \qquad (46)$$

In this equation, β is the initial expansion rate and τ_0 is the characteristic catalyst lifetime.

18.4   SIMULTION OF CNT'S MECHANICAL PROPERTIES

18.4.1   MODELING TECHNIQUES

The theoretical efforts in simulation of CNT mechanical properties can be categorized in three groups as follow:
- Atomistic simulation
- Continuum simulation
- Nano-scale continuum modeling

18.4.1.1   ATOMISTIC SIMULATION

Based on interactive forces and boundary conditions, atomistic modeling predicts the positions of atoms. Atomistic modeling techniques can be classified into three main categories, namely the molecular dynamics, Monte Carlo (MC) and ab initio approaches. Other atomistic modeling techniques such as tight bonding molecular dynamics (TBMD, local density (LD), density functional theory (DFT, Morse potential function model, and modified Morse potential function model were also applied as discussed in last session.

The first technique used for simulating the behaviors of CNTs was molecular dynamic method. This method uses realistic force fields (many-body interatomic potential functions) to determination the total energy of a system of particles. Whit the calculation of the total potential energy and force fields of a system, the realistic calculations of the behavior and the properties of a system of atoms and molecules can be acquired. Although the main aspect of both MD and MC simulations methods is based on second Newton's law, MD methods are deterministic approaches, in comparison to the MC methods that are stochastic ones.

In spite the MD and MC methods depend on the potentials that the forces acting on atoms by differentiating inter atomic potential functions, the ab initio techniques are accurate methods which are based on an accurate solution of the Schrödinger equation. Furthermore the ab initio techniques are potential-free methods wherein the atoms forces are determined by electronic structure calculations progressively.

In generally, MD simulations provide good predictions of the mechanical properties of CNTs under external forces. However, MD simulations take long times to produce the results and consumes a large amount of computational resources, especially when dealing with long and multi-walled CNTs incorporating a large number of atoms.

18.4.1.2   CONTINUUM MODELING

Continuum mechanics-based models are used by many researches to investigate properties of CNTs. The basic assumption in these theories is that a CNT can be modeled as a continuum structure which has continuous distributions of mass, density, stiffness, etc. So, the lattice structure of

CNT is simply neglected in and it is replaced with a continuum medium. It is important to meticulously investigate the validity of continuum mechanics approaches for modeling CNTs, which the real discrete nano-structure of CNT is replaced with a continuum one. The continuum modeling can be either accomplished analytical (micromechanics) or numerically representing FEM.

Continuum shell models used to study the CNT properties and showed similarities between MD simulations of macroscopic shell model. Because of the neglecting the discrete nature of the CNT geometry in this method, it has shown that mechanical properties of CNTs were strongly dependent on atomic structure of the tubes and like the curvature and chirality effects, the mechanical behavior of CNTs cannot be calculated in an isotropic shell model. Different from common *shell model*, which is constructed as an isotropic continuum shell with constant elastic properties for SWCNTs, the MBASM model can predict the chirality induced anisotropic effects on some mechanical behaviors of CNTs by incorporating molecular and continuum mechanics solutions. One of the other theory is shallow shell theories, this theory are not accurate for CNT analysis because of CNT is a nonshallow structure. Only more complex shell is capable of reproducing the results of MD simulations.

Some parameters, such as wall thickness of CNTs are not well defined in the continuum mechanics. For instance, value of 0.34 nm, which is interplanar spacing between graphene sheets in graphite is widely used for tube thickness in many continuum models.

The finite element method works as the numerical methods for determining the energy minimizing displacement fields, while atomistic analysis is used to determine the energy of a given configuration. This is in contrast to normal finite element approaches, where the constitutive input is made via phenomenological models. The method is successful in capturing the structure and energetic of dislocations. Finite element modeling is directed by using 3D beam element, which is as equivalent beam to construct the CNT. The obtained results will be useful in realizing interactions between the nanostructures and substrates and also designing composites systems.

18.4.1.3   NANO-SCALE CONTINUUM MODELING

Unlike to continuum modeling of CNTs where the entirely discrete structure of CNT is replaced with a continuum medium, nano-scale continuum modeling provides a rationally acceptable compromise in the modeling process by replacing C–C bond with a continuum element. In the other hand, in nano-scale continuum modeling the molecular interactions between C–C bonds are captured using structural members whose properties are obtained by atomistic modeling. Development of nano-scale continuum theories has stimulated more excitement by incorporating continuum mechanics theories at the scale of nano. Nano-scale continuum modeling is usually accomplished numerically in the form of finite element modeling. Different elements consisting of rod, truss, spring and beam are used to simulate C–C bonds. The two common method of nano-scale continuum are quasi-continuum and equivalent-continuum methods, which have been used in nano-scale continuum modeling. The *quasi-continuum (QC)* method, which presents a relationship between the deformations of a continuum with that of its crystal lattice, uses the classical Cauchy–Born rule and representative atoms. The quasi-continuum method, mixes atomistic-continuum formulation and is based on a finite element discretization of a continuum mechanics variation principle.

The *equivalent continuum method* developed by providing a correlation between computational chemistry and continuum structural mechanics. It has considered being equal total molecular potential energy of a nanostructure with the strain energy of its equivalent continuum elements.

This method has been proposed for developing structure-property relationships of nano-structured materials and works as a link between computational chemistry and solid mechanics by substituting discrete molecular structures with equivalent-continuum models. It has been shown that this substitution may be accomplished by equating the molecular potential energy of a nano-structured material with the strain energy of representative truss and continuum models. Because of the approach uses the energy terms that are associated with molecular mechanics modeling, a brief description of molecular mechanics is given first followed by an outline of the equivalent-truss and equivalent-continuum model development.

18.4.1.3.1 MOLECULAR MECHANICS

An important part in molecular mechanics calculations of the nano-structure materials is the determination of the forces between individual atoms. This description is characterized by a force field. In the most general form, the total molecular potential energy, E, for a nano-structured material is described by the sum of many individual energy contributions:

$$E_{total} = \sum E_{\rho} + \sum E_{\theta} + \sum E_{\tau} + \sum E_{\omega} + \sum E_{vdw} + \sum E_{el} \qquad (47)$$

where $E\rho$, $E\theta$, $E\tau$, $E\omega$ are the energies associated with bond stretching, angle variation, torsion, and inversion, respectively. The atomic deformation mechanisms are illustrated in Figs. 18.5 and 18.6.

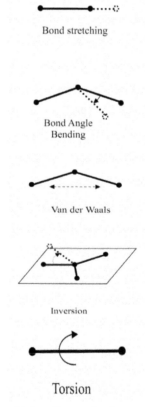

FIGURE 18.5   Atomistic bond interaction mechanisms.

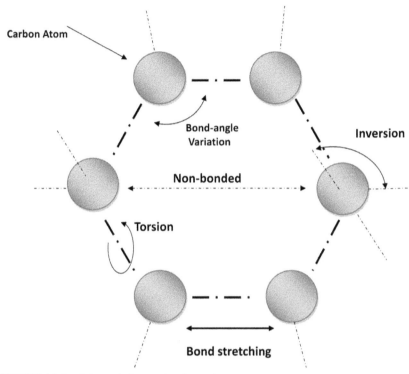

FIGURE 18.6   Schematic interaction for carbon atoms.

The nonbonded interaction energies consist of van der Waals E_{vdw}, and electrostatic E_{el}, terms. Depending on the type of material and loading conditions various functional forms may be used for these energy terms. In condition where experimental data are either unavailable or very difficult to measure, quantum mechanical calculations can be a source of information for defining the force field.

In order to simplify the calculation of the total molecular potential energy of complex molecular structures and loading conditions, the molecular model substituted by intermediate model with a pin-jointed Truss model based on the nature of molecular force fields, to represent the energies given by Eq. (4.1), where each truss member represents the forces between two atoms.

So, a truss model allows the mechanical behavior of the nano-structured system to be accurately modeled in terms of displacements of the atoms. This mechanical representation of the lattice behavior serves as an intermediate step in linking the molecular potential with an equivalent-continuum model.

In the truss model, each truss element represented a chemical bond or a nonbonded interaction. The stretching potential of each bond corresponds with the stretching of the corresponding truss element. Atoms in a lattice have been viewed as masses that are held in place with atomic forces that is similar to elastic springs. Therefore, bending of Truss elements is not needed to simulate the chemical bonds, and it is assumed that each truss joint is pinned, not fixed, Fig. 18.7 shown the atomistic-based continuum modeling and RVE of the chemical, truss and continuum models.

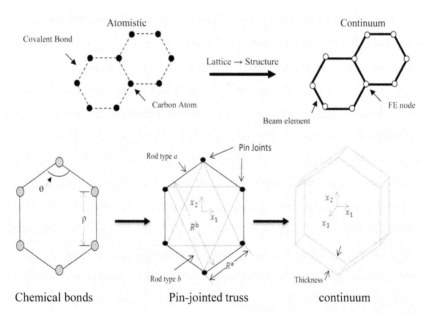

FIGURE 18.7   Atomistic-based continuum modeling and RVE of the chemical, truss and continuum models.

The mechanical strain energy, Λ^t, of the truss model is expressed in the form:

$$\Lambda^t = \sum_n \sum_m \frac{A_m^n Y_m^n}{2 R_m^n} (r_m^n - R_m^n)^2 \tag{48}$$

where A_m^n and Y_m^n are the cross-sectional area and Young's modulus, respectively, of rod m of truss member type n. The term is the stretching of rod m of truss member type n, where R_m^n and r_m^n are the undeformed and deformed lengths of the truss elements, respectively.

In order to represent the chemical behavior with the truss model, Eq. (47) must be consider being equal with Eq. (48) in a physically meaningful manner. Each of the two equations is a sum of energies for particular degrees of freedom. In comparison of the Eq. (47) that has bond angle variance and torsion terms the Eq. (48) has stretching term only, so it made the main difficulty for substitute these two equations. No generalization can be made for overcoming this difficulty for every nano-structured system. It means that possible solution must be determined for a specific nano-Structured material depending on the geometry, loading conditions, and degree of accuracy sought in the model.

18.5   LITERATURE REVIEW ON CNT SIMULATION

Different researchers have been doing many efforts to simulate mechanical properties of CNT, in generally the main trends of these methods employed by different researchers to predict the elastic modulus of SWCNTs results in terms of three main parameters of morphology: radius, chirality, and wall thickness. The dependency of results to the diameter of CNT becomes less pronounced when nonlinear inter atomic potentials are employed instead of linear ones.

There are plenty of experimental and theoretical techniques for characterizing Young's modulus of carbon nanotubes, all three main group of CNT modeling; molecular dynamics (MD), continuum modeling (CM) and nano-scale continuum modeling (NCM) have been used in the literature. Some of recent different theoretical methods for predicting Young's modulus of carbon nanotubes are summarized in Tables 18.1–18.3.

TABLE 18.1   MD Methods for Prediction Young' Modulus of CNT

Researchers	Year	Method	Young's modulus (TPa)	Results
Liew et al. [1]	2004	Second generation reactive of empirical bond-order (REBO)	1.043	Examining the elastic and plastic properties of CNTs under axial tension

TABLE 18.1   *(Continued)*

Researchers	Year	Method	Young's modulus (TPa)	Results
H. W. Zhang [2]	2005	Modified Morse Potentials and Tersoff–Brenner potential	1.08	Predicting the elastic properties of SWCNTs based on the classical Cauchy–Born rule
Agrawal et al. [3]	2006	A combination of a second generation reactive empirical bond order potential and vdW interactions	0.55, 0.73, 0.74, 0.76	Predicting Young's modulus by four MD approaches for an armchair (14,14) type SWCNT and investigating effect of defects in the form of vacancies, van der Waals (vdW) interactions, chirality, and diameter
Cheng et al. [4]	2008	MD simulations using Tersoff– Brenner potential to simulate covalent bonds while using Lennard–Jones to model interlayer interactions	1.4 for armchair 1.2 for zigzag	Evaluating the influence of surface effect resulting in relaxed unstrained deformation and in-layer nonbonded interactions using atomistic continuum modeling approach
Cai et al. [5]	2009	Linear scaling self-consistent charge, density functional tight binding (SCC-DFTB) and an ab initio Dmol3	---	Investigating the energy and Young's modulus as a function of tube length for (10.0) SWCNTs
Ranjbartoreh and Wang [6]	2010	Large-scale atomic/ molecular massively parallel simulator (LAMMPS) code	0.788 for armchair 1.176 for zigzag	Effects of chirality and Van der Waals interaction on Young's modulus, elastic compressive modulus, bending, tensile, compressive stiffness, and critical axial force of DWCNTs

TABLE 18.2   Continuum simulation methods for prediction Young' Modulus of CNT

Researchers	Year	Method	Young's modulus (TPa)	Results
Sears and Batra [7]	2004	Equivalent continuum tube	2.52	Results of the molecular-mechanics simulations of a SWNT is used to derive the thickness and values of the two elastic modulus of an isotropic linear elastic cylindrical tube equivalent to the SWNT
Wang [8]	2004	Stretching and rotating springs, Equivalent continuum plate	0.2–0.8 for zigzags 0.57–0.54 for arm-chairs	Effective in-plane stiffness and bending rigidity of SWCNTs
Kalamkarov et al. [9]	2006	Asymptotic homogenization, cylindrical network shell	1.71	Young's modulus of SWCNTs
Gupta and Batra [10]	2008	Equivalent continuum structure	0.964 ± 0.035	Predicting Young's modulus of SWCNTs b equating frequencies of axial and torsional modes of vibration of the ECS to those of SWCNT computed through numerical simulations using the MM3 potential
Giannopoulos et al. [11]	2008	FEM, spring elements	1.2478	Young's and Shear modulus of SWCNTs using three dimensional spring-like elements
Papanikos et al. [12]	2008	3-D beam Element	0.4–2.08	3D FE analysis, assuming a linear behavior of the C–C bonds

TABLE 18.3   Nano-scale continuum methods for prediction Young' Modulus of CNT

Researchers	Year	Method	Young's modulus (TPa)	Results
Li and Chou [13]	2003	Computational model using beam element and nonlinear truss rod element	1.05 ± 0.05	Studying of elastic behavior of MWCNTs and investigating the influence of diameter, chirality and the number of tube layers on elastic modulus
Natsuki et al. [14]	2004	Frame structure model with a spring constant for axial deformation and bending of C–C bond using analytical method	0.61–0.48	A two dimensional continuum-shell model which is composed of the discrete molecular structures linked by the carbon–carbon bonds
Natsuki and Endo [15]	2004	Analytical analysis by exerting nonlinear Morse potential	0.94 for armchair (10,10)	Mechanical properties of armchair and zigzag CNT are investigated. The results show the atomic structure of CNT has a remarkable effect on stress– strain behavior
Xiao et al. [16]	2005	Analytical molecular structural mechanics based on the Morse potential	1–1.2	Young's modulus of SWCNTs under tension and torsion and results are sensitive to the tube diameter and the helicity
Tserpes and Papanikos [17]	2005	FE method employing 3D elastic beam element	For chirality (8,8) and thickness 0.147 nm 2.377	Evaluating Young's modulus of SWCNT using FEM and investigating the influence of wall thickness, tube diameter and chirality on CNT Young's modulus

TABLE 18.3 *(Continued)*

Researchers	Year	Method	Young's modulus (TPa)	Results
Jalalahmadi and Naghdabadi [18]	2007	Finite element modeling employing beam element based on Morse potential	3.296, 3.312, 3.514	Predicting Young's modulus using FEM and Morse potential to obtain mechanical properties of beam elements, Moreover investigating of wall thickness, diameter and chirality effects on Young's modulus of SWCNT
Meo and Rossi [19]	2007	Nonlinear and torsion springs	0.912 for zigzag structures 0.920 for armchair structures	Predicting the ultimate strength and strain of SWCNTs and effects of chirality and defections
Pour Akbar Safari et al. [20]	2008	Finite element method using 3D beam element	1.01 for (10,10)	Obtaining Young modulus of CNT to use it in FE analysis and investigating Young modulus of CNT reinforced composite
Cheng et al. [21]	2008	MD simulations using Tersoff-Brenner potential to simulate covalent bonds while using Lennard–Jones to model interlayer interactions. Finite element analysis employing nonlinear spring element and equivalent beam element to model in-layer nonbonded interactions and covalent bond of two neighbor atoms, respectively	1.4 for armchair and 1.2 for zigzag	Evaluating the influence of surface effect resulting in relaxed unstrained deformation and in-layer nonbonded interactions using atomistic continuum modeling approach

TABLE 18.3   *(Continued)*

Researchers	Year	Method	Young's modulus (TPa)	Results
Avila and Lacerda [22]	2008	FEM using 3D beam element to simulate C–C Bond	0.95–5.5 by altering the CNT radius Constructing SWCNT (armchair, zigzag and chiral)	Model based on molecular mechanic and evaluating its Young's modulus and Poisson's ratio
Wernik and Meguid [23]	2010	FEM using beam element to model the stretching component of potential and also nonlinear rotational spring to take account the angle-bending component	0.9448 for armchair (9, 9)	Investigating nonlinear response of CNT using modified Morse potential, also studying the fracture process under tensile loading and torsional bucking
Shokrieh and Rafiee [24]	2010	Linking between interatomic potential energies of lattice molecular structure and strain energies of equivalent discrete frame structure	1.033–1.042	A closed-form solution for prediction of Young's modulus of a graphene sheet and CNTs and finite element modeling of CNT using beam element
Lu and Hu [25]	2012	Using nonlinear potential to simulate C–C bond with considering elliptical-like cross section area of C–C bond	For diameter range 0.375–1.8, obtained Young's modulus is 0.989–1.058	Predicting mechanical properties of CNT using FEM, also investigating rolling energy per atom to roll graphene sheet to SWNT
Rafiee and Heidarhaei [26]	2012	Nonlinear FEM using nonlinear springs for both bond stretching and bond angle variations	1.325	Young's modulus of SWCNTs and investigating effects of chirality and diameter on that

18.5.1 BUCKLING BEHAVIOR OF CNTS

One of the technical applications on CNT related to buckling properties is the ability of nanotubes to recover from elastic buckling, which allows them to be used several times without damage. One of the most effective parameters on buckling behaviors of CNT on compression and torsion is chirality that thoroughly investigated by many researchers [27, 28]. Chang et al. [29] showed that zigzag chirality is more stable than armchair one with the same diameter under axial compression. Wang et al. [30] reviewed buckling behavior of CNTs recently based on special characteristics of buckling behavior of CNTs.

The MD simulations have been used widely for modeling buckling behavior of CNTs [31–35]. MD use simulation package which included Newtonian equations of motion based on Tersoff–Brenner potential of inter atomic forces. Yakobson et al. [36] used MD simulations for buckling of SWCNTs and showed that CNTs can provide extreme strain without permanent deformation or atomic rearrangement. Molecular dynamics simulation by destabilizing load that composed of axial compression, torsion have been used to buckling and postbuckling analysis of SWCNT [37].

The buckling properties and corresponding mode shapes were studied under different rotational and axial displacement rates which indicated the strongly dependency of critical loads and buckling deformations to these displacement rates. Also buckling responses of multiwalled carbon nanotubes associated to torsion springs in electromechanical devices were obtained by Jeong [38] using classical MD simulations.

Shell theories and molecular structural simulations have been used by some researchers to study buckling behavior and CNT changeability for compression and torsion. Silvestre et al. [39] showed the inability of Donnell shell theory [40] and shown that the Sanders shell theory [41] is accurate in reproducing buckling strains and mode shapes of axially compressed CNTs with small aspect ratios. It's pointed out that the main reason for the incorrectness of Donnell shell model is the inadequate kinematic hypotheses underlying it.

It is exhibited that using Donnell shell and uniform helix deflected shape of CNT simultaneously, leads to incorrect value of the critical angle of twist, conversely Sanders shell model with nonuniform helix deflected shape presents correct results of critical twist angle. Besides, Silvestre et al. [42] presented an investigation on linear buckling and postbuckling behavior of CNTs using molecular dynamic simulation under pure shortening and twisting.

Ghavamian and Öchsner [43] were analyzed the effect of defects on the buckling behavior of single and multiwalled CNT based on FE method considering three most likely atomic defects including impurities, vacancies (carbon vacancy) and introduced disturbance. The results demonstrate that the existence of any type kinks in CNTs structure, conducts to lower critical load and lower buckling properties.

Zhang et al. [44] performed an effort on the accuracy of the Euler Bernoulli beam model and Donnell shell model and their nonlocal counterparts in predicting the buckling strains of single-walled CNTs. Comparing with MD simulation results, they concluded that the simple Euler–Bernoulli beam model is sufficient for predicting the buckling strains of CNTs with large aspect ratios (i.e., length to diameter ratio $L/d > 10$). The refined Timoshenko's beam model for nonlocal beam theory is needed for CNTs with intermediate aspect ratios (i.e., $8 < L/d < 10$). The Donnell thin shell theory is unable to capture the length dependent critical strains obtained by MD simulations for CNTs with small aspect ratios (i.e., $L/d < 8$) and hence this simple shell theory is unable to model small aspect ratio CNTs. Tables 18.4–18.6 summarizes some of theoretical simulations of buckling behavior of CNTs.

TABLE 18.4 MD Simulation of CNT Buckling

Researchers	Year	Method	Chirality	Length (nm)	Diameter (nm)	Results
Wang et al. [30]	2005	Tersoff–Brenner Potential	(n,n)(n,0)	7–19	0.5–1.7	Obtaining critical stresses and comparing between the buckling behavior in nano and macroscopic scale
Sears and Batra [45]	2006	MM3 class II pair wise potential, FEM, equivalent continuum structures using Euler buckling theory	Various zigzag CNTs	50–350Å	---	Buckling of axially compressed multiwalled carbon nanotubes by using molecular mechanics simulations and developing continuum structures equivalent to the nanotubes
Xin et al. [46]	2007	Molecular dynamics simulation using Morse potential, harmonic angle potential and Lennard–Jones potential	(7,7) and (12,0)	---	---	Studying buckling behavior of SWCNTs under axial compression based on MD method, also evaluating the impression of tube length, temperature and configuration of initial defects on mechanical properties of SWCNTs
Silvestre et al. [39]	2011	The second generation reactive empirical bond order (REBO) potential	(5,5)(7,7)	---	---	Comparison between buckling behavior of SWCNTs with Donnell and Sanders shell theories and MD results
Ansari et al. [35]	2011	"AIREBO" potential	(8,8)(14,0)	---	---	Axial buckling response of SWCNTs based on nonlocal elasticity continuum with different BCs and extracting nonlocal
Silvestre et al. [47]	2012	Tresoff-Bernner Covalent Potential	(8,8) (5,5) (6,3)	---	---	Buckling behavior of the CNTs under Pure shortening and Pure twisting as well as their precritical and postcritical stiffness

TABLE 18.5 Continuum Simulation of CNT Buckling

Researchers	Year	Method	Chirality	Length (nm)	Diameter (nm)	Results
He et al. [48]	2005	Continuum cylindrical shell	---			Establishing an algorithm for buckling analysis of multiwalled CNTs based on derived formula which considering Van der Waals interaction between any two layers of MWCNT
Ghorbanpour Arani et al. [49]	2007	FEM, cylindrical shell	---	11.2 and 44.8	3.2	Pure axially compressed buckling and combined loading effects on SWCNT
Yao et al. [50]	2008	FEM, elastic shell theory	---	---	0.5–3	Bending buckling of single- double-and multi walled CNTs
Guo et al. [51]	2008	Atomic scale finite element method	(15,0) (10,0)	8.3 and 8.38	---	Bending buckling of SWCNTs
Chan et al. [52]	2011	Utilizing Donnell shell equilibrium equation and also Euler–Bernoulli beam equation incorporating curvature effect	Various chirality	---	---	Investigating pre and postbuckling behavior of MWCNTs and multiwalled carbon nano peapods considering vdW interactions between the adjacent walls of the CNTs and the interactions between the fullerenes and the inner wall of the nanotube
Silvestre [53]	2012	Donnell and Sanders shell model with non-uniform helix deflected shape		Various length to diameter ratio	Various length to diameter ratio	Investigating of critical twist angle of SWCNT and comparing the accuracy of two Donnell and Sanders shell model with uniform and nonuniform helix deflected shape

TABLE 18.6. Nano-scale Continuum Simulation of CNT Buckling

Researchers	Year	Method	Chirality	Length (nm)	Diameter (nm)	Results
Li and Chou [54]	2004	Molecular structural mechanics with 3D space frame-like structures with beams	(3,3)(8,8) (5,0)(14,0)	---	0.4–1.2	Buckling behaviors under either compression or bending for both single-and double-walled CNTs
Chang et al. [29]	2005	Analytical molecular mechanics, potential energy	---	---	1–5	Critical buckling strain under axial compression
Hu et al. [55]	2007	Molecular structural mechanics with 3D beam elements	Various type of armchair and zigzag	Various length to diameter ratio	Various length to diameter ratio	Investigating of buckling characteristics of SWCNT and DWCNT by using the beam element to model C–C bond and proposed rod element to model vdW forces in MWCNT, also the validity of Euler's beam buckling theory and shell buckling mode are studied

18.5.2   VIBRATIONS ANALYSIS

Dynamic mechanical behaviors of CNTs are of importance in various applications, such as high frequency oscillators and sensors [56]. By adding CNTs to polymer the fundamental frequencies of CNT reinforced polymer can be improved remarkably without significant change in the mass density of material [57]. It is importance indicate that the dynamic mechanical analysis confirms strong influence of CNTs on the composite damping properties [58].

The simulation methods of CNT's vibrating were reviewed by Gibson et al. [59] in 2007. Considering wide applications of CNTs, receiving natural frequencies and mode shapes by assembling accurate theoretical model. For instance, the oscillation frequency is a key property of the resonator when CNTs are used as nano mechanical resonators. Moreover, by using accurate theoretical model to acquire natural frequencies and mode shapes, the elastic modulus of CNTs can be calculated indirectly.

Timoshenko's beam model used by Wang et al. [60] for free vibrations of MWCNTs study; it was shown that the frequencies are significantly over predicted by the Euler's beam theory when the aspect ratios are small and when considering high vibration modes. They indicated that the Timoshenko's beam model should be used for a better prediction of the frequencies especially when small aspect ratio and high vibration modes are considered.

Hu et al. presented a review of recent studies on continuum models and MD simulations of CNT's vibrations briefly [61]. Three constructed model of SWCNT consisting of Timoshenko's beam, Euler–Bernoulli's beam and MD simulations are investigated and results show that fundamental frequency decreases as the length of a SWCNT increases and also the Timoshenko's beam model provides a better prediction of short CNT's frequencies than that of Euler–Bernoulli beam's model. Comparing the fundamental frequency results of transverse vibrations of cantilevered SWCNTs it can be seen that both beam models are not able to predict the fundamental frequency of cantilevered SWCNTs shorter than 3.5 nm.

An atomistic modeling technique and molecular structural mechanics are used by Li and Chou [62] to calculated fundamental natural frequency of SWCNTs. A free and forced vibrations of SWCNT have been assessed by Arghavan and Singh [63] using space frame elements with extensional, bending and torsional stiffness properties to modeling SWCNTs and compared their results with reported results by of Sakhaee pour [64] and Li and Chou [62]. Their results were in close agreement with two other results, within three to five percent.

Furthermore, free vibrations of SWCNTs have been studied by Gupta et al. [65] using the MM3 potential. They mentioned that calculations based on modeling SWCNTs as a beam will over estimate fundamental frequencies of the SWCNT. They derived the thickness of a SWCNT/ shell to compare the frequencies of a SWCNT obtained by MM simulations with that of a shell model and provided an expression for the wall thickness in terms of the tube radius and the bond length in the initial relaxed configuration of a SWCNT. Aydogdu [66] showed that axial vibration frequencies of SWCNT embedded in an elastic medium highly over estimated by the classical rod model because of ignoring the effect of small length scale and developed an elastic rod model based on local and nonlocal rod theories to investigate the small scale effect on the axial vibrations of SWCNTs.

The vibration properties of two and three junctioned of carbon nanotubes considering different boundary conditions and geometries were studied by Fakhrabadi et al. [67]. The results show that the tighter boundaries, larger diameters and shorter lengths lead to higher natural frequencies. Lee and Lee [68] fulfilled modal analysis of SWCNT's and nanocones (SWCNCs) using finite element method with ANSYS commercial package. The vibration behaviors of SWCNT with fixed beam and cantilever boundary conditions with different cross-section types consisting of circle and ellipse were constructed using 3D elastic beams and point masses. The nonlinear vibration of an embedded CNT's were studied by Fu et al. [69] and the results reveal that the nonlinear free vibration of nano-tubes is effected significantly by surrounding elastic medium.

The same investigation has been accomplished by Ansari and Hemmatnezhad [70] using the variational iteration method (VIM). Wang et al. [71] have studied axis symmetric vibrations of SWCNT immerged in water which in contrast to solid-liquid system, a submerged SWCNT is coupled with surrounding water via vdW interaction. The analysis of DWCNTs vibration characteristics considering simply support boundary condition are carried out by Natsuki et al. [72] based on Euler–Bernoulli's beam theory. Subsequently, it was found that the vibration modes of DWCNTs are noncoaxial intertube vibrations and deflection of inner and outer nanotube can come about in the same or opposite deflections.

More-over the vibration analysis of MWCNTs were implemented by Aydogdu [73] using generalized shear deformation beam theory (GSD-BT). Parabolic shear deformation theory (PSDT) was used in the specific solutions and the results showed remarkable difference between PSDT and Euler beam theory and also the importance of vdW force presence for small inner radius. Lei et al. [74] have presented a theoretical vibration analysis of the radial breathing mode (RBM) of DWCNTs subjected to pressure based on elastic continuum model. It was shown that the frequency of RBM increases perspicuously as the pressure increases under different conditions.

The influences of shear deformation, boundary conditions and vdW coefficient on the transverse vibration of MWCNT were studied by Ambrosini and Borbón [75, 76]. The results reveal that the noncoaxial intertube frequencies are independent of the shear deformation and the boundary conditions and also are strongly influenced by the vdW coefficient used. The performed investigations on the simulation of vibrations properties classified by the method of modeling are presented in Tables 18.7–18.9. As it can be seen 5–7, 5–8, and 5–9, the majority of investigations used continuum modeling and simply replaced a CNT with hollow thin cylinder to study the vibrations of CNT. This modeling strategy cannot simulate the real behavior of CNT, since the lattice structure is neglected. In other word, these investigations simply studied the vibration behavior of a continuum cylinder with equivalent mechanical properties of CNT. Actually nano-scale continuum modeling is preferred for investigating vibrations of CNTs, since it was reported that natural frequencies of CNTs depend on both chirality and boundary conditions.

TABLE 18.7 MD Methods for Prediction Vibrational Properties of CNT

Researchers	Year	Method	Chirality	Aspect ratio (L/D)	Results
Gupta and Batra [77]	2008	MM3 potential	Nineteen armchair, zigzag, and chiral SWCNTs have been discussed	15	Axial, torsion and radial breathing mode (RBM) vibrations of free–free unstressed SW-CNTs and identifying equivalent continuum structure
Gupta et al. [65]	2010	MM3 potential	Thirty-three armchair, zigzag band chiral SWCNTs have been discussed	3–15	Free vibrations of free end SWCNTs and identifying equivalent continuum structure of those SWCNTs
Ansari et al. [78]	2012	Adaptive Intermolecular Reactive Empirical Bond Order (AIREBO) potential	(8,8)	8.3–39.1	Vibration characteristics and comparison between different gradient theories as well as different beam assumptions in predicting the free vibrations of SWCNTs
Ansari et al. [79]	2012	Tersoff–Brenner and Lennard–Jones potential	(5,5)(10,10)(9,0)	3.61–14.46	Vibrations of single- and double-walled carbon nanotubes under various layer-wise boundary condition

TABLE 18.8 Continuum Methods for Prediction Vibration Properties of CNTs

Researchers	Year	Method	Chirality	Aspect ratio (L/D)	Results
Wang et al. [60]	2006	Timoshenko beam theory	0	10, 30, 50, 10	Free vibrations of MWCNTs using Timoshenko beam
Sun and Liu [80]	2007	Donnell's equilibrium equation	---	---	Vibration characteristics of MWCNTs with initial axial loading
Ke et al. [81]	2009	Eringen's nonlocal elasticity theory and von Karman geometric nonlinearity using nonlocal Timoshenko beam	---	10, 20, 30, 40	Nonlinear free vibration of embedded double-walled CNTs at different boundary condition
Ghavanloo and Fazelzadeh [82]	2012	Anisotropic elastic shell using Flugge shell theory	Twenty-four armchair, zigzag and chiral SWCNTs have been discussed	---	Investigating of free and forced vibration of SWCNTs including chirality effect
Aydogdu Metin [66]	2012	Nonlocal elasticity theory, an elastic rod model	---	---	Axial vibration of single walled carbon nanotube embedded in an elastic medium and investigating effect of various parameters like stiffness of elastic medium, boundary conditions and nonlocal parameters on the axial vibration
Khosrozadeh and Hajabasi[83]	2012	Nonlocal Euler–Bernoulli beam theory	---	---	Studying of nonlinear free vibration of DWCANTs considering nonlinear interlayer Van der Waals interactions, also discussing about nonlocal vibration

TABLE 18.9 Nano-scale Continuum Modeling Simulation for Prediction Vibration Properties of CNTs

Researchers	Year	Method	Chirality	Aspect ratio (L/D)	Results
Georgantzinos et al. [84]	2008	Numerical analysis based on atomistic microstructure of nanotube by using linear spring element	Armchair and zigzag	Various aspect ratio	Investigating vibration analysis of single-walled carbon nanotubes, considering different support conditions and defects
Georgantzinos and Anifantis [85]	2009	FEM, based on linear nano-springs to simulate the interatomic behavior	Armchair and zigzag chiralities	3–20	Obtaining mode shapes and natural frequencies of MWCNTs as well as investigating the influence of van der Waals interactions on vibration characteristics at different BCs
Sakhaee Pour et al. [64]	2009	FEM using 3D beam element and point mass	Zigzag and armchair	Various aspect ratio	Computing natural frequencies of SWCNT using atomistic simulation approach with considering bridge and cantilever like boundary conditions

18.5.3   SUMMARY AND CONCLUSION ON CNT SIMULATION

The CNTs modeling techniques can be classified into three main categories of atomistic modeling, continuum modeling and nano-scale continuum modeling. The atomistic modeling consists of molecular dynamic (MD), Monte Carlo (MC) and ab initio methods. Both MD and MC methods are constructed on the basis of second Newton's law. While MD method deals with deterministic equations, MC is a stochastic approach. Although both MD and MC depend on potential, ab initio is an accurate and potential free method relying on solving Schrödinger equation. Atomistic modeling techniques are suffering from some shortcomings which can be summarized as: (I) inapplicability of modeling large number of atoms (II) huge amount of computational tasks (III) complex formulations. Other atomistic methods such as tight bonding molecular dynamic, local density, density functional theory, Morse potential model and modified Morse potential model are also available which are in need of intensive calculations.

On the other hands, continuum modeling originated from continuum mechanics are also applied to study mechanical behavior of CNTs. Comprising of analytical and numerical approaches, the validity of continuum modeling has to be carefully observed wherein lattice structure of CNT is replaced with a continuum medium. Numerical continuum modeling is accomplished through finite element modeling using shell or curved plate elements. The degree to which this strategy, that is, neglecting lattice structure of CNT, will lead us to sufficiently accurate results is under question. Moreover, it was extensively observed that almost all properties of CNTs (mechanical, buckling, vibrations and thermal properties) depend on the chirality of CNT; thus continuum modeling cannot address this important issue.

Recently, nano-scale continuum mechanics methods are developed as an efficient way of modeling CNT. These modeling techniques are not computationally intensive like atomistic modeling and thus they are able to be applied to more complex system with-out limitation of short time and/or length scales. Moreover, the discrete nature of the CNT lattice structure is kept in the modeling by replacing C–C bonds with a continuum element. Since the continuum modeling is employed at the scale of nano, therefore the modeling is called as nanoscale continuum modeling.

The performed investigations in literature addressing mechanical properties, buckling, vibrations and thermal behavior of CNT are reviewed and classified on the basis of three aforementioned modeling techniques. While atomistic modeling is a reasonable modeling technique for this purpose, its applicability is limited to the small systems. On the other hand, the continuum modeling neglects the discrete structure of CNT leading to inaccurate results.

Nano-scale continuum modeling can be considered as an accept-able compromise in the modeling presenting results in a close agreement with than that of atomistic modeling. Employing FEM as a computationally powerful tool in nanoscale continuum modeling, the influence of CNT chirality, diameter, thickness and other involved parameters can be evaluated conveniently in comparison with other methods. Concerning CNTs buckling properties, many researchers have conducted the simulation using MD methods. The shell theories and molecular mechanic structure simulation are also applied to assess the CNTs buckling in order to avoid time consuming simulations. But, for the specific case of buckling behavior, it is not recommended to scarify the lattice structure of CNT for less time consuming computations. But instead, nanoscale continuum modeling is preferred.

Comparing the results with the obtained results of MD simulation, it can be inferred from literature that for large aspect ratios (i.e., length to diameter ratio $L/d > 10$) the simple Euler–Bernoulli beam is reliable to predict the buckling strains of CNTs while the refined Timoshenko's beam model or their nonlocal counterparts theory is needed for CNTs with intermediate aspect ratios (i.e., $8 < L/d < 10$). The Donnell thin shell theory is incapable to capture the length dependent critical strains for CNTs with small aspect ratios (i.e. $L/d < 8$). On the other hand, Sanders shell theory is accurate in predicting buckling strains and mode shapes of axially compressed CNTs with small aspect ratios. From the dynamic analysis point of view, the replacement of CNT with a hollow cylinder has to be extremely avoided, despite the widely employed method. In other word, replacing CNT with a hollow cylinder will not only lead us to inaccurate results, but also there will not be any difference between and nano-structure in the form of tube with continuum level of modeling. It is a great importance to keep the lattice structure of the CNT in the modeling, since the discrete structure of CNT play an important role in the dynamic analysis. From the vibration point of view, MD simulations are more reliable and NCM approaches are preferred to CM techniques using beam elements.

KEYWORDS

- **Carbon Nanotubes**
- **Computational Chemistry**
- **Computational Mechanics**

REFERENCES

1. Liew, K. M., He, X. Q., & Wong, C. W. (2004). On the study of elastic and plastic properties of multiwalled carbon nanotubes under axial tension using molecular dynamics simulation. Acta Mater; 52, 2521–2527.
2. Zhang, H. W., Wang, J. B., & Guo, X., (2005). Predicting the elastic properties of single-walled carbon nanotubes, J Mech Phys Solids; 53, 1929–1950.
3. Agrawal Paras, M., Sudalayandi Bala, S., Raff Lionel, M., & Komanduri Ranga. (2006). A comparison of different methods of Young's modulus determination for single-wall carbon nanotubes (SWCNT) using molecular dynamics (MD) simulations, Comput Mater Sci. 38, 271–281.
4. Cheng Hsien-Chie, Liu Yang-Lun, Hsu Yu-Chen, & Chen Wen-Hwa. (2009). Atomistic-continuum modeling for mechanical properties of single-walled carbon nanotubes, Int J Solids Struct; 46, 1695–1704.
5. Cai, J., Wang, Y. D., & Wang, C. Y. (2009). Effect of ending surface on energy and Young's modulus of single-walled carbon nanotubes studied using linear scaling quantum mechanical method. Physical B; 404, 3930–3934.
6. Ranjbartoreh Ali Reza., & Wang Guoxiu. (2010). Molecular dynamic investigation of mechanical properties of armchair and zigzag double-walled carbon nanotubes under various loading conditions. Phys Lett A, 374, 969–974.
7. Sears, A., & Batra, R. C. (2004). Macroscopic properties of carbon nanotubes from molecular-mechanics simulations. Phys Rev B; 69, 235–406.
8. Wang, Q. (2004). Effective in-plane stiffness and bending rigidity of armchair and zigzag carbon nanotubes. Int J Solids Struct 41, 5451–5461.
9. Kalamkarov, A. L., Georgiades, A. V., Rokkam, S. K., Veedu, V. P., & Ghasemi Nejhad, M. N. (2006). Analytical and numerical techniques to predict carbon nanotubes properties Int J Solids Struct; 43, 6832–6854.
10. Gupta, S. S., & Batra, R. C. (2008). Continuum structures equivalent in normal mode vibrations to single-walled carbon nanotubes, Comput Mater Sci 43, 715–723.
11. Giannopoulos, G. I., Kakavas, P. A., & Anifantis, N. K. (2008). Evaluation of the effective mechanical properties of single-walled carbon nanotubes using a spring based finite element approach. Comput Mater Sci; 41(4), 561–569.

12. Papanikos, P., Nikolopoulos, D. D., & Tserpes, K. I. (2008). Equivalent beams for carbon nanotubes. Comput Mater Sci. 43, 345–352.
13. Li, Chunyu., Chou, & Tsu-Wei. (2003). Elastic moduli of multiwalled carbon nanotubes and the effect of van der Waals forces, Compos Sci. Technol. 63, 1517–1524.
14. Natsuki, T., Tantrakarn, K., & Endo, M. (2004). Prediction of elastic properties for single-walled carbon nanotubes, Carbon; 42, 39–45.
15. Natsuki Toshiaki, & Endo Morinobu. (2004). Stress simulation of carbon nanotubes in tension and compression Carbon; 42, 2147–2151.
16. Xiao, J. R., Gama, B. A., Gillespie, & Jr. J. W., (2005). An analytical molecular structural mechanics model for the mechanical properties of carbon nanotubes, Int J Solids Struct; 42, 3075–3092.
17. Tserpes, K. I., & Papanikos, P. (2005). Finite element modeling of single-walled carbon nanotubes. Composites: Part B; 36, 468–477.
18. Jalalahmadi, B., & Naghdabadi, R. (2007). Finite element modeling of single-walled carbon nanotubes with introducing a new wall thickness, J Phys: Conf Ser 61, 497–502.
19. Meo, M., & Rossi, M. (2006). Prediction of Young's modulus of single wall carbon nanotubes by molecular mechanics based finite element modeling. Compos Sci. Technol. 66, 1597–605.
20. PourAkbar Saffar Kaveh., JamilPour Nima., Najafi Ahmad Raeisi., Rouhi Gholamreza., Arshi Ahmad Reza., & Fereidoon Abdolhossein, et al. (2008). A finite element model for estimating Young's modulus of carbon nanotube reinforced composites incorporating elastic cross-links. World Acad. Sci. Eng. Technol. 47.
21. Cheng Hsien-Chie., Liu Yang-Lun., Hsu Yu-Chen., & Chen Wen-Hwa. (2009). Atomistic-continuum modeling for mechanical properties of single-walled carbon nanotubes, Int J Solids Struct 46, 1695–1704.
22. Ávila Antonio Ferreira., Lacerda Guilherme Silveira Rachid. (2008). Molecular mechanics applied to single-walled carbon nanotubes. Mater Res; 11(3), 325–333.
23. Wernik Jacob, M., & Meguid Shaker, A. (2010). Atomistic-based continuum modeling of the nonlinear behavior of carbon nanotubes, Acta Mech. 212, 167–179.
24. Shokrieh Mahmood, M., & Rafiee Roham. (2010). Prediction of Young's modulus of graphene sheets and carbon nanotubes using nanoscale continuum mechanics approach, Mater Des; 31, 790–795.
25. Lu Xiaoxing., & Zhong, Hu. (2012). Mechanical property evaluation of single-walled carbon nanotubes by finite element modeling, Composites Part B; 43, 1902–1913.
26. Rafiee Roham., & Heidarhaei Meghdad. (2012). Investigation of chirality and diameter effects on the Young's modulus of carbon nanotubes using nonlinear potentials, Compos Struct, 94, 2460–2464.
27. Zhang, Y. Y., Tan, V. B. C., & Wang, C. M. (2006). Effect of chirality on buckling behavior of single-walled carbon nanotubes, J Appl Phys. 100, 074304.
28. Chang, T. (2007). Torsional behavior of chiral single-walled carbon nanotubes is loading direction dependent. Appl Phys Letter 90, 201910.
29. Chang Tienchong., Li Guoqiang., & Xingming, Guo. (2005). Elastic axial buckling of carbon nanotubes via a molecular mechanics model, Carbon; 43, 287–294.
30. Wang, C. M., Zhang, Y. Y., Xiang, Y., & Reddy, J. N. (2010). Recent studies on buckling of carbon nanotubes. Appl. Mech. Rev; 63, 030804.

31. Srivastava, D., Menon, M., & Cho, K. J. (1999). Nanoplasticity of single-wall carbon nanotubes under uniaxial compression Phys Rev Letter 83(15), 2973–2976.

32. Ni, B., Sinnott, S. B., Mikulski, P. T, et al. (2002). Compression of carbon nanotubes filled with C–60, CH4, or Ne: predictions from molecular dynamics simulations. Phys Rev Letter 88, 205–505.

33. Wang, Yu., Wang Xiu-xi., Ni Xiang-gui., & Wu Heng-an. (2005). Simulation of the elastic response and the buckling modes of single-walled carbon nanotubes, Comput Mater Sci; 32, 141–6.

34. Hao Xin, Qiang Han., & Xiaohu Yao. (2008). Buckling of defective single-walled and double-walled carbon nanotubes under axial compression by molecular dynamics simulation, Compos Sci Technol; 68, 1809–1814.

35. Ansari, R., Sahmani, S., & Rouhi, H. (2011). Rayleigh–Ritz axial buckling analysis of single-walled carbon nanotubes with different boundary conditions, Phys Letter A, 375, 1255–1263.

36. Yakobson, B. I., Brabec, C. J., & Bernholc, J. (1996). Nanomechanics of carbon tubes: instabilities beyond linear response. Phys Rev Lett; 76, 2511–2514.

37. Zhang Chen-Li., & Shen Hui-Shen. (2006). Buckling and postbuckling analysis of single-walled carbon nanotubes in thermal environments via molecular dynamics simulation, Carbon; 44, 2608–2616.

38. Jeong Byeong, Woo, & Sinnott Susan, B. (2010). Unique buckling responses of multiwalled carbon nanotubes incorporated as torsion springs. Carbon; 48, 1697–1701.

39. Silvestre, N., Wang, C. M., Zhang, Y. Y., & Xiang, Y. (2011). Sanders shell model for buckling of single-walled carbon nanotubes with small aspect ratio. Compos Struct; 93, 1683–1691.

40. Donnell, L. H. (1933). Stability of thin-walled tubes under torsion, NACA Report No. 479.

41. Sanders, J. L. (1963). Non-linear theories for thin shells, Quart J Appl. Math; 21, 21–36.

42. Silvestre Nuno, Faria Bruno., Canongia Lopes, & José, N. A. (2012). Molecular dynamics study on the thickness and postcritical strength of carbon nanotubes. Compos Struct; 94, 1352–1358.

43. Ghavamian Ali, & Öchsner Andreas. (2012). Numerical investigation on the influence of defects on the buckling behavior of single-and multiwalled carbon nanotubes, Physica, E; 46, 241–249.

44. Zhang, Y. Y., Wang, C. M., Duan, W. H., Xiang, Y., & Zong, Z. (2009). Assessment of continuum mechanics models in predicting buckling strains of single-walled carbon nanotubes. Nanotechnology; 20, 395–707.

45. Sears, A., & Batra, R. C., (2006). Buckling of multiwalled carbon nanotubes under axial compression. Phys Rev B; 73, 085410.

46. Xin Hao., Han Qiang., & Yao Xiao-Hu. (2007). Buckling and axially compressive properties of perfect and defective single-walled carbon nanotubes, Carbon; 45, 2486–2495.

47. Silvestre Nuno., Faria Bruno., Canongia Lopes, & José, N. (2012). A molecular dynamics study on the thickness and postcritical strength of carbon nanotubes, Compos Struct; 94, 1352–1358.

48. He, X. Q., Kitipornchai, S., & Liew, K. M. (2005). Buckling Analysis of Multi-Walled Carbon Nanotubes: A Continuum Model Accounting For Van Der Waals Interaction, J Mech Phys Solid; 53, 303–326.
49. Ghorbanpour Arani, A., Rahmani, R., & Arefmanesh, A. (2008). Elastic buckling analysis of single-walled carbon nanotube under combined loading by using the ANSYS software, Physical E; 40, 2390–2395.
50. Yao, Xiaohu., Han Qiang., & Xin Hao. (2008). Bending buckling behaviors of single and multiwalled carbon nanotubes. Comput. Mater Sci. 43, 579–590.
51. Guo, X., Leung, A. Y. T., He, X. Q., Jiang, H., & Huang, Y. (2008). Bending buckling of single-walled carbon nanotubes by atomic-scale finite element, Composites: Part B; 39, 202–208.
52. Chan Yue., Thamwattana Ngamta., Hill James, M. (2011). Axial buckling of multiwalled carbon nanotubes and nanopeapods Eur J Mech A Solid; 30, 794–806.
53. Silvestre Nuno. (2012). On the accuracy of shell models for torsional buckling of carbon nanotubes, Eur J Mech.A/Solid; 32, 103–108.
54. Chunyu, Li., & Chou Tsu-Wei. (2004). Modeling of elastic buckling of carbon nanotubes by molecular structural mechanics approach, Mech. Mater; 36, 1047–1055.
55. Hu, N., Nunoya, K., Pan, D., Okabe, T., & Fukunaga, H. (2007). Prediction of buckling characteristics of carbon nanotubes. Int J Solids Struct; 44, 6535–6550.
56. Sawano, S., Arie, T., & Akita, S. (2010). Carbon nanotube resonator in liquid. Nano Lett; 10, 3395–3398.
57. Formica Giovanni., Lacarbonara Walter., & Alessi Roberto. (2010). Vibrations of carbon nanotube-reinforced composites, J Sound Vib; 329, 1875–1889.
58. Khan Shafi Ullah., Li Chi Yin., Siddiqui Naveed, A., & Kim Jang-Kyo. (2011). Vibration damping characteristics of carbon fiber-reinforced composites containing multiwalled carbon nanotubes, Compos Sci. Technol. 71, 1486–1494.
59. Gibson, R. F., Ayorinde, E. O., & Wen, Y. F. (2007). Vibrations of carbon nanotubes and their composites: a review. Compos Sci. Technol. 67, 1–28.
60. Wang, C. M., Tan, V. B. C., & Zhang, Y. Y. (2006). Timoshenko beammodel for vibration analysis of multiwalled carbon nanotubes, J Sound Vib; 294, 1060–1072.
61. Hu, Yan-Gao., Liew, K. M., & Wang, Q. (2012). Modeling of vibrations of carbon nanotubes, Proc Eng 31, 343–347.
62. Li., C., & Chou, T. W. (2003). Single-walled carbon nanotubes as ultrahigh frequency nanomechanical resonators, Phys Rev B; 68, 073405.
63. Arghavan, S., & Singh, A. V. (2011). On the vibrations of single-walled carbon nanotubes, J Sound Vib; 330, 3102–3122.
64. Sakhaee-Pour, A., Ahmadian, M. T., & Vafai, A. (2009). Vibrational analysis of single-walled carbon nanotubes using beam element. Thin Wall Struct; 47, 646–652.
65. Gupta, S. S., Bosco, F. G., & Batra, R. C. (2010). Wall thickness and elastic moduli of single-walled carbon nanotubes from frequencies of axial, torsional and in extensional modes of vibration, Comput Mater Sci; 47, 1049–1059.
66. Aydogdu, Metin. (2012). Axial vibration analysis of nanorods (carbon nanotubes) embedded in an elastic medium using nonlocal elasticity. Mech Res Commun; 43, 34–40.
67. Fakhrabadi Mir Masoud Seyyed., Amini, Ali., Rastgoo Abbas. (2012). Vibrational properties of two and three junctioned carbon nanotubes, Comput Mater Sci; 65, 411–425.

68. Lee, J. H., Lee, B. S. (2012). Modal analysis of carbon nanotubes and nanocones using FEM. Comput Mater Sci. 51, 30–42.
69. Fu, Y. M., Hong, J. W., & Wang, X. Q. (2006). Analysis of nonlinear vibration for embedded carbon nanotubes. J Sound Vib 296, 746–756.
70. Ansari, R., & Hemmatnezhad, M. (2011). Nonlinear vibrations of embedded multiwalled carbon nanotubes using a variational approach, Math Comput Modell; 53, 927–938.
71. Wang, C. Y., Li, C. F., & Adhikari, S. (2010). Axisymmetric vibration of single-walled carbon nanotubes in water, Phys Lett A; 374, 2467–2474.
72. Natsuki Toshiaki, Ni., Qing-Qing., & Endo Morinobu. (2008). Analysis of the vibration characteristics of double-walled carbon nanotubes, Carbon; 46, 1570–1573.
73. Aydogdu Metin. (2008). Vibration of multiwalled carbon nanotubes by generalized shear deformation theory, Int J Mech Sci; 50, 837–844.
74. Lei Xiao-Wen., Natsuki Toshiaki., Shi, Jin-Xing., Ni, & Qing-Qing. (2011). Radial breathing vibration of double-walled carbon nanotubes subjected to pressure, Phys Lett A; 375, 2416–2421.
75. Ambrosini Daniel., & Borbón Fernanda, De. (2012). on the influence of the shear deformation and boundary conditions on the transverse vibration of multiwalled carbon nanotubes, Comput Mater Sci; 53, 214–219.
76. Borbón Fernanda, De., & Ambrosini Daniel. (2012). On the influence of van der Waals coefficient on the transverse vibration of double walled carbon nanotubes, Comput Mater Sci. 65, 504–508.
77. Gupta, S. S., & Batra, R. C. (2008). Continuum structures equivalent in normal mode vibrations to single-walled carbon nanotubes. Comput Mater Sci. 43. 715–723.
78. Ansari, R., Gholami, R., & Rouhi, H. (2012). Vibration analysis of single-walled carbon nanotubes using different gradient elasticity theories, Composites: Part B; 43(8), 2985–2989.
79. Ansari, R., Ajori, S., & Arash, B. (2012). Vibrations of single and double-walled carbon nanotubes with layer-wise boundary conditions: a molecular dynamics study. Curr Appl Phys; 12, 707–711.
80. Sun, C., & Liu, K. (2007). Vibration of multiwalled carbon nanotubes with initial axial loading. Solid State Commun. 143, 202–207.
81. Ke, L. L., Xiang, Y., Yang, J., & Kitipornchai, S. (2009). Nonlinear free vibration of embedded double-walled carbon nanotubes based on nonlocal Timoshenko beam theory. Comput Mater Sci. 47, 409–417.
82. Ghavanloo, E., & Fazelzadeh, S. A. (2012). Vibration characteristics of single-walled carbon nanotubes based on an anisotropic elastic shell model including chirality effect. Appl Math Model; 36, 4988–5000.
83. Khosrozadeh, A., & Hajabasi, M. A. (2012). Free vibration of embedded double-walled carbon nanotubes considering nonlinear interlayer van der Waals forces, Appl Math Model, 36, 997–1007.
84. Georgantzinos, S. K., Giannopoulos, G. I., & Anifantis, N. K. (2009). An efficient numerical model for vibration analysis of single-walled carbon nanotubes Comput Mech. 43, 731–41.

85. Georgantzinos, S. K., & Anifantis, N. K. (2009). Vibration analysis of multiwalled carbon nanotubes using a spring–mass based finite element model. Comput Mater Sci. 47, 168–177.

INDEX

Milton Keynes UK
Ingram Content Group UK Ltd.
UKHW022103141024
449569UK00031B/1759

9 781774 633465